经济管理学术文库·其他类

羊毛毡材料技术的传承问题研究

Wool felt handicraft's inheritance research

宝·阿茹娜／著

图书在版编目（CIP）数据

羊毛毡材料技术的传承问题研究/宝·阿茹娜著 . —北京：经济管理出版社，2019.5
ISBN 978 - 7 - 5096 - 6443 - 8

Ⅰ. ①羊… Ⅱ. ①宝… Ⅲ. ①羊毛—毛毡—手工艺品—材料技术—研究 Ⅳ. ①TS973.5

中国版本图书馆 CIP 数据核字（2019）第 050579 号

组稿编辑：曹　靖
责任编辑：张巧梅
责任印制：黄章平
责任校对：陈　颖

出版发行：经济管理出版社
　　　　　（北京市海淀区北蜂窝 8 号中雅大厦 A 座 11 层　100038）
网　　址：www.E - mp.com.cn
电　　话：(010) 51915602
印　　刷：北京玺诚印务有限公司
经　　销：新华书店
开　　本：720mm×1000mm/16
印　　张：7.75
字　　数：127 千字
版　　次：2019 年 5 月第 1 版　2019 年 5 月第 1 次印刷
书　　号：ISBN 978 - 7 - 5096 - 6443 - 8
定　　价：68.00 元

·版权所有　翻印必究·
凡购本社图书，如有印装错误，由本社读者服务部负责调换。
联系地址：北京阜外月坛北小街 2 号
电话：(010) 68022974　　邮编：100836

前　言

　　羊毛毡材料是游牧民族上千年流传至今的不可或缺的传统生活材料，而羊毛毡手工艺制作不仅是游牧民族的传统生产技能之一，也是人类重要的手工艺文化之一。但由于现代工业技术发展和人们生活习惯的变化致使这一重要文化逐渐被人遗忘。因此，为了保护和发展这些民族文化，中国政府从 21 世纪初开始在全国范围内大力推进传统手工艺复兴运动，寻找并赋予这些传统文化更多的价值和深层含义。在上述背景下，本书以内蒙古地区的蒙古族羊毛毡手工艺为主要研究对象，对天然羊毛毡材料的历史、制作过程、材料的显性价值和隐性价值等问题进行了系统性的实地调研及梳理和分析，并从现代生活习惯和传统文化含义的双重视角探讨了传统天然羊毛毡的深层文化内涵和现代设计领域的广泛利用价值。

INTRODUCTION

To learn more about ancient wool felt and conduct further research in this field, I have chosen the Inner Mongolia Autonomous Region of China as my research location. The reason for this choice is first and foremost that this is my home and growth region and that I hope to find a job there in the future. Secondly, Inner Mongolia is an autonomous region dominated by Mongolian ethnic groups among the 55 ethnic minorities in China whose nomadic language and culture have been perpetuated for thousands of years. It has the largest grassland in China, a very large livestock sector and very rich wool resources; it is one of the main production areas for Chinese wool. Wool materials play an important role in the history of nomadic culture and wool felt craftsmanship is one of the oldest inventions of nomads. I believe that nomadic meadows of Inner Mongolia, as a region of origin of ancient materials, have a little – known heritage that needs to be developed, both in terms of wool felt and in terms of understanding nomadic culture.

One of the oldest textile materials, wool felt, has undergone many transformations and evolutions. In Inner Mongolia, in the 40 years since the founding of the People's Republic of China in 1949, the wool textile industry has made a significant contribution to the country. Its development made it possible to solve the problem of clothing, food and life for hundreds of millions of people at that time. Many large state – owned wool textile companies were established in the Hohhot region of Inner Mongolia. However, the rapid development of the economy caused Chinese state – owned wool textile companies to collapse in the 1990s. The main reason was errors in market management such as the prevalence of unscrupulous traders, the employment of unskilled workers, the development of illegal private individual workshops that diverted branded wool products, manufactured counterfeit and poor quality products in the main public and private companies at the time of the formation of the famous wool battle. More serious was the shift of the raw wool and cashmere market from Inner Mongolia (Grassland) to the industrial province of Hebei, leading to an underground "black market". The black market was widespread and the incorporation of wool into the sand destroyed the quality of the products. This commodity war has had a huge impact on large public companies. Later, the state formulated new policies and many large wool textile companies were forced into "bankruptcy and reorganization", but this could not prevent the fate of counterfeiting by other private companies. China's industrial economic development has brought to the market more and more new materials that have slowly replaced the original wool. Chemical or synthetic fibres at low cost and with the advantages of high manufacturing efficiency have gradually taken over the entire market. Economic development has disrupted society, more and more people have changed their li-

INTRODUCTION

festyles and the nomadic population has decreased significantly.

Today, more and more people are inheriting and learning the craft of wool felt, especially nomads who enter urban life from nomadic areas. Some craft associations in Inner Mongolia have brought together craftsmen to make wool felt and independent craftsmen have set up their own wool manufacturing workshops. As I noticed that more and more people were responding to the "slogan" of craftsmanship and joining this vigorous movement, I began to think: in the movement of rebirth of Inner Mongolia, can we find a new role for these soon – to – be – forgotten materials?

With these questions in mind, I started working to find information. According to data, wool felt is one of the oldest materials, it is not a material reserved only for nomadic populations, it spread from nomadic areas all over the world thousands of years ago, especially in some cold regions of northern Europe. After the Industrial Revolution, since the period when artistic movements were born in Europe, a large number of craft techniques have been revived and these movements are at the origin of the history of modern design. Nowadays, over time, wool felt has become one of the materials used by many modern European artists and designers and appearing in many fields such as product design, furniture design, etc. So, what is the development of wool felt in the nomadic grassland area where its raw materials are produced? I spent my vacation time doing a preliminary market research on wool felt products in Hohhot, the capital of the Inner Mongolia Autonomous Region of China. I visited 129 Mongolian product shops, Mongolian Commodity City, as well as the Mongolian International Exhibition and found that the proportion of wool felt products was very low, about 2% and that most belonged to the most expensive handicrafts and paintings category, with practical products being rare. There are not many books on wool felt material in this region and only a few museums can present them. Since China has implemented the rebirth movement of craftsmen, there have been many felt craftsmen in Inner Mongolia. I interviewed three craftspeople in this industry in different regions, including a teacher, an independent craftsman and a business leader. I asked a few questions to try to understand their point of view on the current state of the material in nomadic areas.

During the interviews with the craftsmen, I first learned the technology of making wool felt. I have a good understanding of the characteristics of wool felt materials that I have divided into two families: the explicit value and the hidden value. I have a detailed knowledge of the three current categories of craftsmen, conditions experienced and needs to date, in the wool craft industry. I have also tried to develop, based on their experiences, the current problems in the nomadic felt area industry for analysis. I think that an experience similar to the rebirth of craftsmanship, a hundred years after that of Europe, could be developed in China. There will still be many problems to solve before this return, but since the development of a series of artistic movements in Europe is a process in which craftsmen have transformed themselves into modern designers, I think that in a similar process of transformation, old wool felt can give more value and possibilities to Mongolian craftsmen. I believe that its return will not only allow us to think more about the natural environment, but also to generate more responsible design values.

ACKNOWLEDGEMENTS

I would like to thank all people who helped me in the preparation of this thesis.

First of all, I would like to thank Mr. Christophe BARDIN, Professor at Jean Monnet University for his teaching and advice throughout my studies.

I also thank the craftsmen Li Qiang, Ulantoyaa and Dalaiqeqeg for answering all my questions with their enthusiasm and patience.

I would particularly like to thank Mrs. Sylvie FOREST and Mr. GEN Suo for their precious help in proofreading and correcting my work.

Finally, I would also like to thank my friends SASSI Nouha and Sherley Pequignot for helping me to find solutions to complete my thesis.

TABLE OF CONTENTS

WOOL FELT ··· 1

- The history of wool felt ··· 1
- The use and spread of wool worldwide ··············· 4
- Definition of wool felt ·· 7
- Production process of wool felt ···························· 9

APPLICATION OF WOOL FELT

　　——Western inclinations ··· 20

- Changes in the heritage relationship between tradition and craftsmanship with modern design ······················· 20
- Application of wool felt in the field of modern art and European modern design ····································· 22

MARKET SURVEY 27

- Interview of an independent artisan: Li Qiang (Chinese) 28
- Interview with the owner: Ulantoyaa (Mongolia) 32
- Interview of Professor Dalaiqeqeg (Mongolia) 35
- Interview analysis 39

CONCLUSION 65

APPENDIX 68

Li Qiang Interview 68

Ulantoyaa Interview 83

Dalaiqeqeg Interview 95

BIBLIOGRAPHY 106

WEBOGRAPHY 108

WOOL FELT

Wool felt is one of the oldest textiles in the world and its discovery knows no borders. From historical origins of the world, it has been available in different versions. From East and West, various cultures also spread many legends on this subject.

The history of wool felt

Oriental historical materials

According to the China News Network of June 9, 2015, mentioned in the Daily Beijing: "The first felt hats were discovered in Louboubo, Loulan, Xinjiang. They date back to the Neolithic periods (about 3880 years ago) and show

how much men have long been used to felt hats. Zhou Chaozhong's official book mentions that during the third dynasty of China, (1046 to 256 BC). Aristocrats wore wool markers made of animal skins. White wool felt hats that existed during the Tang dynasty (618 to 907 AD) were called "Bai Ti", their shapes were composed of triangles, tops, edges, buckles, etc. Zhang Wei, the writer of the Ming dynasty (1368 to 1644 AD) noted in his ancient encyclopedia: "In the ancient Qin dynasty and the Han dynasty, the technology of nomads making wool felt was introduced in mainland China. Then it became popular in folklore." In the ancient Chinese city of Beijing, there were also many felt hat shops; Beijing, located in northern China where the winter is cold and the wind is blowing, so sales of felt hats were very high.

Western historical Materials

British writer and artist Mary Elizabeth Burkett mentions in her book [1] that the first felt craft objects found were found on Neolithic frescoes (6500 – 3000 BC) in present-day Turkey. The curved decorative wall motif can be reminiscent of the use of felt. The author believes that Neolithic artists accurately copied the wool felt tapestry motifs from the wall paintings. These felt blankets can come from the Bronze or Iron Ages. She mentioned that the first documentary evidence from archaeologists came from China, around 2300. In particular, many wool felt fragments have been found in the wild regions of Central Asia. The most fa-

---

[1] Mary Elizabeth Burkett, 《An Early Date for the Origin of Felt》, December 1977.

mous felt fragments come from the 7th to 2nd century BC, they were discovered in a frozen stone house buried in Pazyryk, in the Altai Mountains, in Southern Siberia, at the beginning of the 20th century. As well as many antiques, they have been kept in ice, isolated from the air. Two large felt murals were found in the tomb of a nomadic chief. These prove that felt wool technology was well developed in the past. They feature rich embroidery motifs and are currently on display at the Hermitage Museum in Russia. Traces of wool felt found in Scandinavia, Sweden can be traced back to the Iron Age.

Oriental legend of the wool felt

There is a legend about felt hats in Shaoxing, Zhejiang, China. In ancient times, a hunter hunted in the mountains. While chasing a tiger, he ran into the cave where the tiger lived. After killing the tiger, the hunter found a "felt" in the den. The tiger would put the rest of the hair in its den after eating pigs, sheep and other animals. Over time, these were crushed and pressed like a pie-shaped felt. Later on the hunter brought it home, washed it and after its transformation it became a hat. It was warm and comfortable, the news spread, then everyone copied the felt hat.

Western legend of the wool felt

There are many legends about wool felt in the West. The most famous of them is the story of Saint Clement and Saint Christopher: Once, Saint Clement ran through the forest to avoid his enemy, his feet began to sore, but he still had to run to get rid of the one who was after him. Suddenly, he found wool in the forest, stopped, started collecting it to relieve the pain in his feet. The collected wool was used to wrap and protect them. When he managed to get rid of the enemy, he took his painful feet out of the shoes and noticed that the wool had become a pair of wool felt shoes. Later, Saint Clement became the saint patron of the manufacturers of shoes and hats made of wool felt. He was often referred to in religious rituals and became a saint who hunted the demon and brought good luck. In the history of the Christian Bible, Noah's Ark, the origin of wool felt has also been mentioned.

The use and spread of wool worldwide

Wool felt is one of the oldest textile materials in human history. It is also at the origin of the oldest crafts of the nomads. In particular, in Central Asia, since prehistoric times, nomadic peoples in Mongolia and Kazakhstan have made extensive use of wool felt materials to make many utensils, such as bed covers, carpets and tent covers. Nomads have called their wool felt habitats "yurts". It is a folding tent with a wooden frame, covered with large markers, some of which are decorated with bright colors. This coating is waterproof, soundproof, durable, foldable, easy to move and transport. Inside the yurt, there are also carpets, cushions, storage accessories and bags, saddles and many other wool felt items. It is also used to make clothing, hats, boots, coats and gloves necessary for nomadic life.

Fig. 01　Mongolian yurt made of wool felt (around 100 years old)

Photo by Bao Aruna 23/02/2018

Ancient Romans and Greeks also used wool felt. Roman soldiers used it as armor to protect themselves against the attack of bows and arrows. After the fall of the Roman Empire, Crusaders brought this material from Constantinople to Europe. Thus, wool felt began to spread from Central Asia to the Mediterranean and the rest of the world. In the 12th and 13th centuries, its use flourished in

Western Europe and then spread to South America. In some cases, he even replaced the leather because it was expensive at the time and not effective against rain and snow; felt boots keep the inside of the shoe dry. During the French Revolution, the Phrygian red felt hat, which symbolizes freedom, began to spread. At the beginning of the 20th century, miners used wool felt helmets to protect their heads. Due to their insulation and waterproofing, Scandinavians and Russians also use wool felt to make thick boots. This day, it is still used by many nomads, especially in difficult climates.

But with the development of industrialization, the manual shaping of felt gradually decreased, especially in the 17th century. However, on the verge of being abandoned, this material has suddenly reappeared in recent years, rediscovered by craftsmen around the world. Designers are also attracted to it, they consider it practical and decorative; they want to explore even more of its possibilities. Some designers mix wool with other natural fibres such as camel hair, silk, cashmere and linen. In the United Kingdom, Scandinavia and the United States, a revival of craftsmanship has developed, including the use of wool felt. Production technology and design from wool felt are becoming more and more numerous and highly sought after by the fashion industry.

After wool felts had been handmade in Europe and America for some time, in Japan and Taiwan there was a trend to design small handicrafts with needle felts. This fun style is very popular and known as the Kawaii style.

In summary, these various developments show that ancient wool felt techniques and materials have been widely used in production and everyday life throughout the world. After a period of decline, it begins to reappear.

Definition of wool felt

Felt is a kind of non-woven wool, composed mainly of sheep and sometimes cattle wool fibres, treated and glued by moistening, rubbing, pressing and friction using the felting properties of the wool fibre. Currently on the market, with the exception of felt made of pure wool fibres, most felts are synthetic fibres made of polyester, viscose, polypropylene, acrylic. In this photo, pure wool felt is the subject of extensive research.

Fig. 02 Wool felt of natural origin

Photo by Bao Aruna 21/02/2018

Type of wool and worldwide distribution
——Wool is one of the most widely used natural fibres

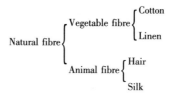

Industrial chain made of wool felt

Industries most directly related to wool felt are livestock, textiles and manufacturing, livestock provides raw materials, wool felt is a kind of textile material – of which we will reveal its way of production in a following age – and is handmade or designed to make wool felt. They constitute an effective cyclical industrial relationship between the three.

Trend in the global wool industry

At present, the global volume of wool exports is unevenly distributed. Australia is the largest wool producing and exporting region. Chinese wool production is the second largest in the world, but China is the largest country in the world in terms of processing. Its volume of wool processing represents 35% of that of the whole world. The main producers and exporters of the global wool textile industry are China, Italy and India. However, China's wool raw materials are mainly dependent on imports, accounting for 80% of wool raw materials from abroad. Chinese wool companies buy wool in two ways. One is purchased directly from an Australian wool supplier and the other is purchased from a local wool factory. In 2010, 80% of Australian wool was exported to China. China is the largest consumer of wool textile products, accounting for 23% of total world consumption. The main European countries such as Italy, Germany, Great Britain and France consume at least 17% of the market, followed by the United States and Japan, which represent 11% and 9% respectively. Data on Chinese wool consumption are important and most of them relate to the clothing industry.

Production process of wool felt

Felts are manufactured in two ways: pure manual work, a single machine and a combination of machine and manual. This material is particularly simple to comb by the manual method. Manual wool felts are roughly divided between wet felt and needle felt.

The felt felting principle of felt humid:

The surface of the wool is covered with tiny scales that we cannot see with naked eyes. When these scales meet hot water, they open and raise. If they are pressed, rubbed by external forces, the scales get tangled up. The wool shrinks, closely amalgamating the fibres together in a felted state, which is the basic principle of "wet felt".

——*Ping guo lu*[①]

1. Traditional hand felting process

The following photo mainly describes the production process of the artist Dalaiqeqeqeg.

Main wet felt materials and tools

Wool, combs, gauze, soap, hot clear water

[①] Ping guo lu, 《Wool Felf》, May 2010.

1.1 Select the wool

First, the wool must be sorted and sieved. In general, the hair of older sheep is harder and the hair of younger sheep is softer. The wool must be sorted according to the quality and type of wool desired. The gross weight and the proportion of the wool thickness are determined according to the expected need.

Fig. 03 – 04 Dalaiqeqeg chooses the wool

Photo by Bao Aruna

1.2 Combing and extracting impurities

A professional wool comb should be used to sort the desired hairs that can remove impurities and the hairs will become softer.

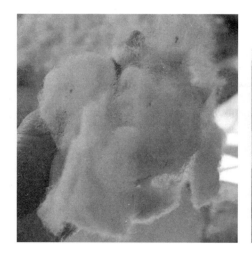

Fig. 05 Combing hairs   Fig. 06 Impurities in the wool

1.3 Paved wool

The combed wool is distributed over the workbench. It is evenly distributed in layers. The number of layers of wool to be laid determines the thickness and size of the final wool felt.

Fig. 07 – 08 Spreading wool

1.4 Cover with gauze

Spread a clean, dry cotton gauze over the wool layer already laid.

Fig. 09 Spreading a gauze

1.5 Wet with water

Spread the clean water evenly over a paved cotton gauze, so that the wool under the gauze completely absorbs moisture, the felting phenomenon occurs.

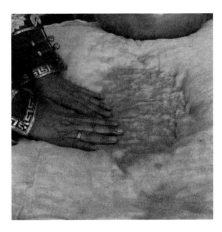

Fig. 10 Wet with water

1.6 Cleaning wool felt

On fully spread gauze, rub with soap with both hands to clean it. Avoid damaging the layered structure of the felt during cleaning and proceed by placing your hands side by side on the gauze in a parallel movement.

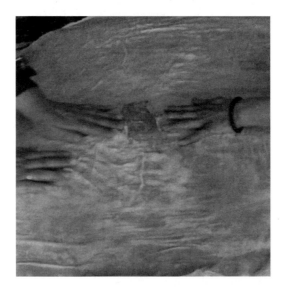

Fig. 11　Cleaning the wool felt

1.7 Wool felt shrinks

Felting reactions occur when wool fibres meet water, as described above. Soapy water completes this shrinking of wool fibres.

WOOL FELT

Fig. 12 Wool felt shrinks under the influence of water and soap

1.8 Rewash the felt and rub it again

After the initial cleaning of the felt, the wool fibres are glued together, the gauze can then be removed. The wool felt thus formed is cleaned again, to perfect its adhesion and waterproofing.

1.9 Pressing the felt several times

The moulded wool felt is pressed several times with both hands. Rub, knead to make it more and more compact and robust and the size of the felt shrinks until the required dimensions are met.

Fig. 13 – 14　Cleaning of wool felt

Fig. 15　Pressing the felt

1.10　Remove impurities

As the finished wool felt is very tight, many impurity particles in the fibres are detected. They become very visible. At this stage, for finishing, a small tool can be used to remove impurities from the wool felt.

Fig. 16　Removing impurities

1.11　Final cleaning of the felt

With the impurities removed, there are still tiny impurities and soap stains left. It must therefore be washed again.

Fig. 17　Final cleaning of the felt

1.12 Spin the felt

The wool felt after cleaning can be spin manually. It may also be placed in the washing machine for more efficient spin drying.

Fig. 18 Spinning of felt

1.13 Wool felt is machine dried or sun dried

After drying the wool felt in the machine, it can be placed on a radiator or dried in the sun naturally.

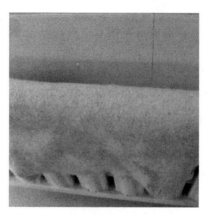

Fig. 19 Drying the wool felt

1.14 processing wool felt

Dry wool felt can be used to design various practical products. Hand – sewn patterns can improve its embellishment.

2. Hand – needled felt

With a special wool needle, pierce the wool several times. This also allows the wool scales to become entangled with each other, thus achieving the felting effect.

——Ping guo lu[1]

---

[1] Ping guo lu, 《Wool Felf》, May 2010.

APPLICATION OF WOOL FELT
——Western inclinations

Changes in the heritage relationship between tradition and craftsmanship with modern design

"I now state, without ambiguity, that the purpose of the applied arts in utilitarian articles is twofold: first, to add beauty to the results of man's work which, if it were true, would be ugly; and second, to add pleasure to the work itself which would otherwise be tedious and discouraging. If this is the case, we must stop being surprised that man has always tried to ornament the work with his hands, that he needs to have around him every day and every hour, or that he has always tried to transform the torments of his labour into pleasure when he thought it possible."

——William Morris[1]

---

[1] William Morris, 《L'Art et l'Artisanat d'Aujourd'hui》, 1889.

At the beginning of the 19th century, industrial revolution affected the whole of Europe, technological progress considerably improved social productive forces and enriched the material life of populations, but it also brought new social problems: from a craft industry for the greatest number, mechanized society led to productivity. This evolution has resulted in both a decline in design quality due to mass production and a proliferation of poorly finished industrial products on the market. In the second half of the 19th century, artists, designers and sociologists, under the emulation of William Morris and John Ruskin, opposed the "no soul" machines induced by the industrial revolution, because they despised these mechanised products produced in series at the time. They initiated a movement to revive art – the **Arts & Crafts movement** in the hope of revitalizing traditional crafts, developing alternatives, improving the aesthetic qualities of everyday objects and rehabilitating handmade work. The Arts & Crafts Movement, also known as the Design Improvement Movement, is at the origin of modern design. In this movement, the emphasis is on the combination of art and technology, the inspiration of nature and the importance of practicality and aesthetic design; they hoped that craftspeople would be able to enjoy their own creations.

Arts and Crafts movement, at its peak, is also consolidated by the production of an additional innovative force through the **New Art** movement. It is an artistic movement, at the end of the 19th and beginning of the 20th century, based on the aesthetics of curved lines. This movement inherited the ideas of Arts & Crafts movement, but it placed more emphasis on naturalism, while being influenced by the elements of oriental art. As the cradle of the "new art", France also had a great interest in Japanese fine arts at that time. At the beginning of the 20th century, under the influence of modernization and industrialization, many artists and designers accepted the inevitability of a new era, no longer avoided mechanical forms and accepted new materials. The industrial manufacture of glass is an expressive field, developed at that time. Movement of the new decorative arts made the mechanical form and modern features more natural and became a new way of innovating at that time. In 1925, the International Exhibition of Decorative Arts was held in Paris and marked the birth of the Decorative **Arts movement**. It is also the development and extension of the Arts & Crafts movement and the crafts of New Art. It has produced a group of highly qualified artists, craftsmen, architects and designers, showing the best of originality. This series of movements led to the creation of modernist designers initiated by the Bauhaus, creating a relationship of heritage and coexistence between traditional craftsmanship and modern industrial design.

Application of wool felt in the field of modern art and European modern design

Wool felt was widely used in the 19*th century and after the First World War. In addition to being used in everyday life, it has also been used in creative design: people use wool felt to tell stories to children and for puppet making. Later, it also appeared in modern art and design.*

APPLICATION OF WOOL FELT

Joseph Beuys (12 may 1921 – 23 january 1986)

He is a German artist famous for his installations and his performance art as his main creative form. His work is varied and many of his creations come from his childhood and his experience of the Second World War. For example, during the war, he was in the Air Force, was seriously wounded and landed on the territory of the Soviet Union, before being found and rescued by local Tatars. Later, German patrol found him and took him to the hospital, but Beuys told them that nomads used a wool felt covered with animal fat and wrapped him around him to save him. Later, these events, which saved his life, reappeared in his artistic creations, such as fat and wool felt. He often used these two materials to commemorate the rescue of his life. He believes that if art is to survive, only one side can be due to God and angels, while the consideration of earth and animals allows us to find a way out. He believes that people should protect nature and be closer to animals. His most famous performance expressing this belief is his 1965 work "**How to explain the paintings to a dead hare**". The spectator can only look through a window: he has poured honey on his head, put a gold leaf and gold powder and nailed an iron plate to his shoe. He holds a dead rabbit in his arms, moving from one painting to another, explaining the painting to the rabbit. From time to time, he stops, holding the dead rabbit, and then crosses the dry branches in the centre of the gallery. This work suggests that he, god, animals and land are linked.

Romero Vallejo

Romero Vallejo is a Spanish designer and architect who works in the fields of architecture, interior design, product design and artistic direction. After graduating in Barcelona, he is currently working with several design institutions and universities of architecture. His design **RUFF POUF** is an attractive furniture accessory, inspired by the fabric ruffles worn around the neck during the Elizabethan era to show wealth and status. Mixing geometry and folds, RUFF pouffe is formed by a central core surrounded by a continuum of folds, all made of 100% natural felt. The combination of hard and soft, abstract and geometric patterns gives this piece of furniture its strong character, making it a bold design object. There are two different sizes, a high and a low pouffe respectively 40 cm and 30 cm. The high version has a diameter of 70 cm while the low version has a diameter of 100 cm higher. It is available in warm colour pairs that combine tones of the inner folds in grey, beige and purple felt for the high version, grey and yellow for the low pouffe.

Metylos

"100% French and eco – responsible manufacturing" is the motto of Design studio Metylos, referenced Ateliers d'Art de France since May 2011. Since the creation of Metylos in Lyon a few years ago, craftsmanship is still considered as the heart of the product developing many unique series of luminaires and small decorations. The two materials used are wood and wool felt. Many furniture designed at Metylos have the signature of a young French designer. Its design philosophy is to minimize the impact of production on environment. Wood comes from sustainably managed forests of Jura, the felt is 100% wool and dyed organically, the assembly is carried out in their workshops. Products are manufactured by many designers who have signed with him, and apply wool felt to fashionable furniture. Products are manufactured according to demand. It is important for Metylos to participate in respecting the environment and for their felt, they have chosen a supplier who uses organic dyes.

Maxime Lapouille

Creator of Tisserand is a French designer Maxime Lapouille. After obtaining his master's degree in industrial design, he is currently working as one of Salomon brand's designers. Tisserand, is a modular carpet to be made by yourself, which can be woven and unweaved at will, using a non – woven material: a 100% natural felt based on 4 strips of different sizes pierced with several slots in their center. This carpet can be extended as required and can be combined freely. It offers unlimited possibilities, and what does not limit the user's imagination is a very interesting design. The felt is fireproof, anti – fouling and non – slip, so it is very suitable for carpeting. This design has passed the daily use test.

Amy Hunting

The Norwegian – born London – based designer Amy Hunting has created a Felt & Gravity, a series of consoles, tables and storage compartments. She uses 100% wool felt and Douglas Fir assembled with solid brass bolts for better fixing. Accessories for this kit – mounted wooden structure are wool felt bands forming shelves under the weight of objects placed inside. The force of gravity becomes one of the main components of this furniture and the 100% wool shelves draw their strength from the weight placed inside.

MARKET SURVEY

    I chose an Inner Mongolian city as the subject of study because it has the largest natural pasture in China. On its meadows, the old nomadic life continues to this day. The original material of felt was the most important material of nomadic culture and everyday life, its development in Inner Mongolia is very typical. Since China is currently in the process of reviving traditional crafts, there are many craftsmen who make wool felt, some have started a business and others are involved in craft associations or local civil government has been founded. I mainly chose three craftsmen from different cities around the meadow of Inner Mongolia who were the main interviewees. Among them are a teacher, an independent craftsman and an entrepreneur. They work mainly in this industry, which often exists in municipalities, and in interviews we mainly used dialogue and communication. We also visited the wool felt manufacturing studios and learned about the complete manufacturing process. We hope that through this research we will be able to present the different types of interviews with craftsmen, analyse similarities and differences in their respective development processes and analyse problems between wool felt and craftsmen as well as the designers' point of view.

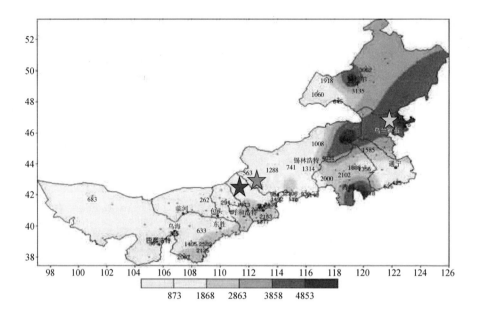

Fig. 20　The map of Inner Mongolia

Interview of an independent artisan:
Li Qiang (Chinese)

Date of Interview: 21/02/2018, 22/02/2018

Location for Interview: Saihantal, Inner Mongolia

MARKET SURVEY

Interview Process: I visited some of the studio's product displays and wool felt machine shops. We conducted a face – to – face interview.

Fig. 21 Li Qiang, un wool felt craftsman

Li Qiang, 46 years. He is an independent Chinese craftsman engaged in this industry for 23 years. His father was an employee, licensed from the wool mill. His studio still had felt machines, recovered from the collapsed wool factories several decades ago. After completing his secondary education at the technical high school, he entered the wool felt industry in 1995 thanks to his father's teaching. He founded his own local studio, workshop and home as a single entity, in the form of family work, without companions. Most of the time, product is manufactured by him independently and in some cases with the help of family members. He said that the current wool felt industry does not have large companies in the region because after the reform of national system, fur factories and factories producing everyday products, especially factories and influential companies in the manufacture of nomadic wool felt, ceased their activities. Following the launch of the revival movement of traditional crafts in China in recent years, he applied to become a technical heir to the felt craft, which gave him more responsibility in the promotion and realization of this trade.

During our conversation, he clarified that his initial idea was to use his own natural wool material and process the product. He refused to add other chemical fibres to the natural wool material. Processes were carried out independently of each other and always by hand. His clients were mainly shepherds on prairies and elderly people in need or patients, as they generally know the equipment and its benefits. Later, after promoting his product on the Internet, other customers asked him for more personalized products. He is very familiar with production and processing of felt, as well as characteristics of the material. He also obtained the certificate of craftsman in wool felt in Inner Mongolia. However, since its material is too primitive compared to current conditions, it is not very familiar with historical source of the felt material. Although some specialists came to perfect his craft skills, I noticed during the interview that there were still many problems.

He thinks that the current material of wool and the development of this profession are very limited. On the one hand, there are fewer and fewer nomads in cold regions and fewer and fewer people who need wool felt objects. Due to current technological development, many new materials have replaced felt and the use of natural wool felt has begun to be abandoned. Although the government has helped them to carry out some promotional activities in the craft revival movement, this has not promoted it much. It is currently in a phase of wait – and – see and ambiguity, hoping that there will be more innovation in the craft or design sector and that it will find a better direction in the future, so as to have more opportunities to change the stagnant state of this old material.

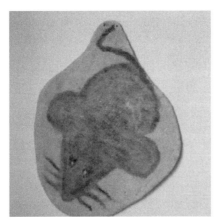

Fig. 22 – 23 Wool felt paint and Mongolian wool felt hat

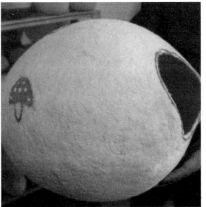

Fig. 24 – 25 Felt boots and wool felt cat room

Interview with the owner: Ulantoyaa (Mongolia)

Date of Interview: 23/02/2018

Location for Interview: Сөнөд баруун хошуу, Inner Mongolia

Interview Process: I visited the craftsmen's studio and discovered the exhibition and use of some wool felt products. We had extensive discussions and exchanges.

Fig. 26　Ulantoyaa, a wool felt craftsman

Ulantoyaa, 50 years. She is not only an independent Mongolian craftsman, but also the owner of a wool felt company. After graduating in Mongolian language from the Normal University of Inner Mongolia, she studied photography and became a professional photographer from 1992 to 1995. Later, after saving some money, she opened a national handicraft shop in Ordos. Her first motivation for entering the wool felt industry was that her father had been a camel officer and that she had always had deep feelings for animals, knowing their habits and characteristics well. In addition, her godmother was also a local woman well known for her craft embroidery. She officially created the felt company in 2010. The company is a family business with two spouses for more than eight years and has more than 100 employees in a government – backed cultural industry at Сөнөд баруун хошуу. The origin of this entrepreneurship is that she believes that wool felt is a material that has been and remains very important, that it is also a craft technology that cannot disappear. Even if, for historical reasons, it has been abandoned for many years, this material must return to people's lives, although many people, especially young people, do not understand and do not know that it is a craft material. She actually believes that this represents a continuity of traditional culture and stresses the importance of the nature of environmental protection. She believes that advantages and functions of wool felt are very numerous, but that they are not sufficiently used in today's life. Although she is very familiar with the material properties of felt, she is not very familiar with the history of wool felt. She only knows what the 70 – years – old Mongolian elder said when he complained about the discovery of wool in cultural relics unearthed 3000 years ago!

During the interview, I learned that since the beginning of her business, she has explored a little bit the process of creating wool felt. She has conducted market studies on wool felt in various fields, but it should be noted that most of the employees of her company are former unemployed people who entered the city from nomadic areas. His company then benefited from an abundant workforce and allowed many people to re – engage in the workforce. In order to promote and increase the quantity of product manufactured, it has also tried to work with other companies, such as yurt manufacturing companies and other interior design companies. It consciously registered the companies' trademarks and requested property rights on the products sold, considering that its attitude was conducive to the promotion and development of the wool felt industry, so that the product could be processed and marketed. His current work consists mainly of buying machine – made wool felt in factories in southern China, then processing and shipping it back into handicrafts. Its felts use natural camel lines and refuse to use chemical fibres such as artificial wool in order to preserve the natural character of the product. She believes that due to the particularity of the grassland area, her company will not be limited by the supply of raw materials and secondly, she will have sufficient human resources. Therefore, she believes that the wool trade can be developed and hopes to find a larger market and more opportunities in the future in the hope that the product will guarantee the conviction that her father passed on to her: nomads live by nature. Everything comes from nature and the nomadic culture that returns to nature can continue to survive thanks to this conviction. It rather works on the spiritual side of culture by aiming to show it in its products.

Fig. 27 – 28 Artisanal teapot and the wool felt cushions

Fig. 29 – 30 Pen holder, chessboard and painting on felt

Interview of Professor Dalaiqeqeg (Mongolia)

Date of Interview: 25/02/2018 – 28/02/2018

Location for Interview: Улаан хот, Inner Mongolia

Interview Process: I visited and learned how to make wool felt, we had a discussion and visited his studio's exhibition of wool felt products.

Fig. 31 Dalaiqeqeg, a wool felt craftsman

Dalaiqeqeg, 55 years. An independent Mongolian craftsman, she obtained a degree in Mongolian studies at the University of Inner Mongolia. After graduation, she worked as a teacher at the Mongolian college from 1983 to 2017. During her teaching years, she used her free time and started making wool felt as a secondary activity for 10 years. When she retired in 2017, the creation of wool felt became her main activity. Most of her work is produced independently and since her daughter also has the necessary skills to work with wool, they sometimes collaborate together. The reason she chose to enter this profession is that her family, and especially her mother, did not go to school in her youth, but had very good craft skills. Several years after her mother's death, she began the process of making felt through her memory and past memories, because she missed her mother too much. She hoped that by rediscovering this ancestral know – how, this skill can be brought to the attention of a greater number of people while advancing the heritage of national culture. During these 10 years of secondary activities, she became more and more interested in wool felt and hopes that this skill can be passed on to others, generating a greater sense of responsibility. Over the past 10 years, she has taught this practice to more than 300 students, but most of them are high school students, so it was mainly a matter of cultivating their interest in this activity. She is very competent and professional on the material properties of felt, but due to the limited information available, she is not particularly competent on the history of felt, knowing only the carpets + from Turkey. To promote this process, she has participated in many local felt competitions and won many awards. She also obtained the certificate of inheritance of the felt craft industry.

During the interview, I learned that her younger brother was a local shepherd and had a large meadow, so the raw materials for wool felt were abundant for her and so she mainly used her own raw materials for wool to process her own products. The two aspects, manufacturing and processing of wool felt, are done independently. However, she is currently starting her business, she is facing a lot of questions and hopes that the product can be more creative and develop on the market. She lacks human resources and considers that her personal production time is too limited. How to build a brand and find a market and how to seek cooperation are her current concerns. She hopes that her products can be used or collected.

Fig. 32 – 33 The Mongolian hat and the needle bag

Fig. 34 – 35 Shoes, pen holder, storage box and soles

Fig. 36 – 37 Wool felt bag and the bow bag

MARKET SURVEY

Interview analysis

On the basis of three interviews with craftsmen, I addressed five themes:

- The value of the wool felt material.
- Handcrafted products made of wool felt.
- Wool felt craftsmen must have better conditions.
- The development of wool felt craftsmanship.
- The current state of the wool felt industry.

1. Value of wool felt material

The three craftsmen interviewed have a rich experience in this sector and a good understanding of the properties of material, This shows that the material itself is highly valued and has real value. Through my interviews with the craftsmen and research carried out in book books on history and material, I understood that the value of wool felt can be analyzed in two ways: **the explicit value and the hidden value**.

The explicit value of the wool felt material, including all its characteristics such as physical properties, chemical properties, functional, structural and geographical advantages, economic advantages, etc., must be taken into account.

The explicit value

Wool felt is insulating: it can keep warm and resist cold or heat

Wool felt has resistance to heat and cold, but it can also insulate against heat. Due to the small diameter of the wool fibre structure and its elasticity, a layer of air is formed in the overlapping fibre layers, which can block cold air from the outside and also block the external heat flow. For example: the Mongolian yurt, in some alpine nomadic areas in China and Mongolia, especially in northern Inner Mongolia, northwestern Xinjiang, the Mongolian plateau, many places where nomadic winter temperatures can reach minus 50 degrees. For a long time, shepherds had been covering the yurt with thick markers to resist the cold. With the stove inside in winter, it is very hot and the contrast is striking. In summer, outside, it is more than thirty degrees and the inside of the yurts is very cool. Physical characteristics of wool are favorable conditions for yurts to be warm in winter and cool in summer. Even during the 13th century war, nomads in winter wore wool felt pens to warm themselves up.

Wool felt and air permeability

Although some wool felts generally appear thick, they are also breathable on the basis of thermal insulation. In the process of overlapping wool fibers, although the organization is tight, there is always a small air circulation in pores, so that many people use it as soles to maintain the permeability of the soles of the feet.

MARKET SURVEY

Fig. 38 Wool carpets inside the yurt

Wool felt and hygroscopicity

Wool fibres knitted in a weft of horizontal and vertical lines. Hygroscopicity is the ability of a material to absorb moisture from the air. When the felt is sprinkled with water, elastic wool fibres completely absorb water, contract and quickly absorb moisture. If the wool felt is dyed, it will receive enough dye and the dyed wool felt will not discolour easily. Compared to other fibres, wool felts have more colour variety due to a better dyeing effect. Machine dyeing is more uniform than manual dyeing. However, the possibility of matching the colours with hand – dyed wool felt will be more diversified. The color gradient effect will be more natural.

Wool felt and water resistance as well as moisture resistance

The structure of the wool fibres are comparable to overlapping tiles to make a waterproof roof. Not only does it absorb moisture, but it also blocks moisture. It has a very strong waterproof characteristic. This is one of the reasons why there are so many cold regions using it. Water droplets float on the surface of the wool felt, the layered structure, formed by the wool fibres, makes the water droplets less likely to penetrate. If there is not much water, simply use a dry cloth to wipe. This is also the reason why wool felt has been used in many regions to prevent rain and snow.

Fig. 39 Waterproof wool felt boots

Wool felt and its sound and acoustic insulation

The density of wool fibres depends on the strength of the wool when it is felted. Due to the repeated combination of high – density and low – density materials, when sound waves move, they pass between materials of different densities. Sound passes through the felt of different density several times; this is the process by which it is absorbed and gradually the sound decreases, until finally effective sound insulation is achieved. Wool fibres have a very good flexibility and during the formation of the felt, its density is easily adjustable. Low density when it is soft, higher density when it is tight, which satisfies the adjustment of noise between low and high frequency. Felt can also effectively reduce noise bouncing and suppress the propagation of various vibrations while reducing echo noise. Felt can be used in combination with other soundproofing materials, such as thin wood panel and cement – based panel. These methods are often found in soundproofing materials for interior decoration. But most of the time, pure wool felt is not used, it is replaced by other chemical fibres or is a combination of fibres. Felts are easy to cut, easy to shape, simple to build and very decorative, they can meet the needs of different styles and categories of soundproofing decoration. It is particularly suitable for places such as cinemas and concert halls where it is essential to absorb noise for certain surfaces. The soundproofing performance of the felt is very high. In some developed European countries, it is of great importance for the interior decoration of the house and other places, however, in China, more promotion will be needed to ensure greater use.

Wool felt and its anti-shock effect

Due to the high flexibility of the wool fibres, it has a good effect in resisting pressure, it can be used for its "cushion" effect. In the process of forming wool felts, adhesion of the material is very high and the organization is tight, therefore when used, it is not easily dispersible when impacted. It is a practical shock-resistant material, of good quality and low cost.

Wool felt and its filtering property

After the felting process, wool fibres become tight, material becomes very thin and the distance between molecules becomes very small and flexible. Since it is a non-woven needle punching process, it behaves differently from weaving. Its thickness is limited when it is felted and is also hygroscopic. Such characteristics not only can easily exclude certain impurities, but dust and dirt can also be filtered. Wool felt is a very good filter material for air purification and indoor filtration.

Wool felt is protective and defensive

Superposed layers of wool fibres can form a compact and robust protective layer; in ancient times, wool felt was used as protective clothing to dampen the damage caused by some sharp weapons. For example, wool felts on the periphery of yurts are also used when the wind blows on the meadows to act as a buffer: sand will be cushioned by the wool without causing injuries to inhabitants.

Wool felt and its friction properties

The surface of the wool fibre is covered with tiny scales. When an external force is applied to push and pull the wool fibres, it produces movements that have friction effects. For example, some tablecloths slide easily on the desk or table, but if they are covered with a layer of wool felt, they will not move easily to avoid slipping.

Wool felt and plasticity

Among natural fibres, wool fibres have the best elasticity and can be easily folded. They are flexible and can lose their elasticity in hot and humid conditions. Under the influence of external forces, they can interpenetrate, shrink and close gradually. After being felted, they can be molded into various shapes and can be kept in shape for a long time under certain conditions. The felt is easy to cut into various sizes to form the object, the edge of the cut will not be loose, in addition, the thickness of wool felt can be completely controlled, but will be of uneven density. At the same time, it also has a good recovery from its original form. When the finished wool felt product is folded, it can quickly restore its original shape and will not easily deform. It's its plastic quality.

Wool felt and fire resistance

The own ignition point of the wool fibre reaches 560 degrees. It is not easily flammable and is used as a safety fibre. Like human hair, it belongs to the category of animal protein fibres. When the wool fibre is burned, layers of carbon dioxide and water – based carbon dioxide are formed. It can effectively prevent the continuous expansion and spread of flame. For example, when a cigarette butt falls on a wool carpet, there is a small hole after combustion that will not continue to spread. Wool felt is used as a fire extinguisher to cover fuels, it can effectively isolate fuel from contact with oxygen in the air. Therefore, wool felt has non – flammable and fire – resistant properties. In the past, many people made carpets out of wool felt. A main characteristic is not only to keep warm but also to prevent combustion.

Wool felt and protection against radiation

Wool is the most resistant fibre among the natural fibres resistant to light and weather. Subjected to 1000 hours of light, its reduction capacity is about 50%, due to ultraviolet light which destroys the disulfide bonds of wool. When the cystine is oxidized, the color gradually turns yellow and the light intensity slowly decreases. Due to its ability to prevent radiation, in some nomadic regions and high latitudes that have a lot of sun, felt will be used instead of sun hats or sun protection clothing.

Wool felt also has a certain role in medical treatment

In many nomadic and craft regions in the past, there was an old medical remedy: people wounded in the meadow, by applying pressure to the wound, using ashes from the blankets after burning, stops the bleeding. This practice would require a real and more reliable medical certification. Many people living in pastoral areas suffer from colds, back pain and leg pain, so it is necessary to

MARKET SURVEY

rely on blanket protection to relieve pain. It is understood that there are also shepherds who use wool felt and sesame oil to cover injured cow horns to prevent frostbite.

Wool felt has characteristics that do not deteriorate easily

Due to its particular structure and material properties, wool felt can be used repeatedly and does not deteriorate easily during long – term storage. Some of the wool rugs have existed in yurts for a few hundred years in the homes of some shepherds and can still be used. The felt exposed in many museums is still well preserved.

Wool felt is a non – toxic and environmentally friendly material

Wool fibres are natural animal fibres, they generate water and carbon dioxide after combustion. It will therefore not produce toxic gases as with the combustion of chemical fibres. It is a renewable and recyclable ecological material. This performance is very important for the daily life of human beings. This material is gradually becoming popular with individuals as an environmentally friendly material. In European and American countries, wool felt has been recognized by some designers, it appears more and more often in the field of fashion clothing.

When choosing a material, the first thing to consider is its explicit value, such as its properties, functions, structural characteristics, etc. , and the hidden value of the material itself is sometimes weakened or ignored. I think that the hidden value of the material, which we often ignore, is perhaps the most moving part of the product and that even this can influence design and its values.

Hidden value

Wool felt materials in nomadic areas have geographical advantages

Inner Mongolia is mainly made up of high plateaus at an altitude of 1000 meters, with the largest and best natural pastures in China covering an area of more than 880000 square kilometers. Production of animal products such as cashmere is the first in China for many years and livestock farming is one of the most important activities in the region. As a place of production of wool materials, it has sufficient geographical advantages and is rich in conditions for the supply of raw materials for wool felt. However, during investigation, artisan Li Qiang stated that much of the wool from the nomadic region of Inner Mongolia is no longer used and that many herders even choose to burn it. This situation concerns not only wool but also some leathers. There are several reasons for this, the first is that for different qualities of wool, Inner Mongolia had the largest base of cashmere production and processing in the world, it was called "The Cashmere Region", and the annual production of cashmere in Inner Mongolia represented 40% of that in the world. Cashmere is the best quality wool and most high quality cashmeres are used in the garment industry. Undesirable rest of the wool or wool of relatively low quality with impurities, has no textile value, some will be transported to the factory in south for the manufacture of parts for industrial use, most of the rest of the wool becomes waste and is abandoned or burnt. Another reason is the change in the social environment of nomadic areas: most herders enter urban life and no longer live on the prairies, resulting in fewer and fewer users. The wool felt used by Ulantoyaa is a machine made wool felt purchased in other major cities. During the development of Inner Mongolia in China, many large felt wool manufacturing companies closed their doors, they did not have necessary conditions for large – scale processing of wool felt machines. But in recent years after China launched the rebirth of craftsmanship, Inner Mongolia gradually began to redo wool felt, produced by a large number of craftsmen thanks to the fact that the grasslands of Inner Mongolia have a geographical advantage and the best conditions to collect the original material.

Green sustainable environmental value

On 12 December 2015, 195 UN Member States adopted the Paris Agreement at the UN Climate Summit. On 22 April 2016, World Earth Day, China signed the Paris Agreement. This agreement has led almost every country in the world, in a community of destiny, to protect the ecology of Earth and to consider the threat of the global environment and climate caused industrial development that has reached an ultimate level. The future society must begin to transform towards green sustainable development. According to CFW, fashion and textile industry is currently the second most polluting industry in the world, generating hundreds of millions of used clothes every year. China has long been known as the world's largest factory, particularly in the fashion and textile industries. Since 1994, production and exports have been ranked first in the world. Faced with such threats of terrible and serious pollution, choice and development of materials have a significant impact on various industries, particularly in the field of industrial design. I have already stated in – the explicit value of wool felt – that it is a natural, non – toxic, primitive, natural material, with renewable durability characteristics. I think that in the face of more and more new chemicals, it is a waste of natural resources that a large quantity of wool is thrown away or burned. After a long period of disappearance of the wool felt material, when, thanks to the rebirth of craftsmanship, it reappears in people's lives, the value of craftsmanship cannot be denied. Involuntarily labelled with the original craftsmanship, it seems to me that these old labels limit to a large extent the development of the material's own value. Obviously, when a craftsman or designer creates and designs a product, it is not enough to know that the material used is environmentally friendly, it is also necessary to understand what it means when they voluntarily choose an environmentally friendly material. For example, in this research, I was interested to learn that one of the disadvantages of wool felt is that insects appear when it is poorly stored; but three craftsmen were unanimous in their resistance to the addition of pesticides to wool felt or other ingredients such as a chemical pigment or other mixed chemical fibres. They are aware of the shortcomings of natural materials, but faced with even more unknown threats, chemical additives, they are determined to maintain the original pure wool. In any case, respect for the environment of materials is the principle to which the three craftsmen adhere and do not want to derogate. We cannot deny the need for scientific progress and future technologies, but I believe that craftsmen and designers have the ability to choose environmentally friendly materials regardless of environmental, economic and social conditions and material production. This ca-

MARKET SURVEY

pacity is more important than the advancement of science and technology, as it will determine the future of humanity. It can determine the meaning of the future product and determine the designer's attitude and design values.

Wool felt has an economic value

When abandoned or destroyed materials can be reintroduced into the lives of users and can be better used, they bring new economic benefits. In this survey, **Dalaiqeqeg** craftsman lived in the town of the nomadic region canton and his brother was a pure local nomad. As a result, it will not suffer much from the economic impact of raw materials in the process of creating its own wool felt products. At the same time, artisan Li Qiang has repeatedly pointed out that wool in his pastoral area is very cheap: about 5 pounds of wool cost only 2 yuan (about 0.2 euros) and many nomads burn a lot of wool. The wool felt used by the craftsman Ulantoyaa is a wool felt processed in the southern factory, but she thinks that the wool from this factory did not pay much to buy its wool materials in Inner Mongolia, processed and resold after processing. This price is still not high, indicating that the initial cost is very low, but the cost of freight is very high. This may explain why the material cost of wool in Inner Mongolia is very low, which is an important advantage in product design, as it will make artisans or designers likely to obtain more economic benefits. Economic value is a favourable condition for stimulating the creative power of more artisans or designers.

Cultural and emotional value

The creation and use of a material represents the civilization and progress of human society over time, each region having its own iconic objects. Until today, wool felt is the oldest fabric in the history of mankind: it is not only a material, it also carries the culture of several millennia, especially for today's nomads. In this survey, among the reasons why the three craftsmen chose and joined this profession, all insisted on the same point: "I hope that thanks to my own insistence, this profession can be transmitted". So, in the revival movement of wool felt craftsmanship, I wondered why they wanted to continue to do so. Was it just to inherit and continue an old material or to know how it is made? In our conversation, the craftsman **Li Qiang** mentioned two things, first of all being satisfied to have customers who come to him for special and personalized products. He was asked to make a hat with the horns of the cow in southern China Miao, because it represents the spiritual Totem ethnic culture, because of the original quality of the wool felt, the plasticity is very strong and it is very easy to transform it into horn. The other thing is that he received an order to make wool felt for an unknown temple. The plaintiff demanded that all wool felts retain their original colour, as it reflects the purity, the most important part of the temple's religious culture. The craftsman **Ulantoyaa** also stated that the nomadic culture represented by wool felt is more significant than the wool felt material itself. For example, she said that there was a skylight on the yurts and that the wool felt in the skylight should use the most "spiritual" and pure sheep's wool possible, that of the lamb. In nomadic culture, the sky represents nature, human beings must respect and surrender to nature, instead of conquering it. On the prairie, nature protection is about protecting the nomads themselves.

MARKET SURVEY

Fig. 40　A skylight of the Mongolian yurt

Photo by Bao Aruna 29/05/2017

This shows that old wool felt sometimes represents an identity spirit, sometimes represents a religious belief and more importantly, it represents the most necessary ecological culture in today's human environment. I think that the cultural ownership of a material is a broader advantage that allows this material to go further: the cultural symbols of the material will give the artisan or designer more confidence and opportunities to create. It will also make the product more moving. A product with an ecological and cultural origin will essentially encourage people to pay attention to the environment. The cultivation of wool felt is also an expression of Mongolian nomadic identity culture, whose return has inspired many Mongolian craftsmen intellectually and to show more solidarity. During the interview, **Ulantoyaa** said she had been exploring the road to felt creation since she started her business and had tried different ways and conducted market research on wool felt in many areas. It should be mentioned that most of the employees of his company are unemployed people who arrive in the city from nomadic areas, which gives him not only an abundant labour force, but also the opportunity to re – engage. In order to promote and create more products, it has also tried to work with other companies, such as yurt and interior design companies. It has consciously registered the company's trademark and requested ownership rights to the products, which is considered conducive to the promotion and development of the wool felt industry, so that the product can be marketed. Its products originate from natural camel lines, rejecting chemical fibres such as artificial wool and preserving the ecological character of the product. It believes that due to the particularity of the grassland area, it will not be limited by the supply of raw materials and will have sufficient human resources. As a result, she believes that her company can prosper and hopes to find a larger market and more opportunities in the future. She also thinks that this company will be able to transmit the idea that the fathers taught her – nomads live by nature and everything they receive comes from nature and will come back from nature, this culture can really continue thanks to wool felt. When the cultural concept is transmitted through the product design, it is significant for the product itself, for the creator and for the user.

2. Handcrafted products made of wool felt

Li Qiang's wool felt products are mainly handicraft products such as felt boots and felt paints and there are fewer functional products. The wool felt products of craftsman **Ulantoyaa** are mainly felt embroideries, cushions, etc., craft and functional products, but always mainly craft products. Craftsman **Dalaiqeqeg**'s wool felt products are mainly wool felt clothing, felt paints and various small products. In addition to clothing, most of them are mainly handicrafts. From the product research process, I learned that those that are only functional are much lower than artisanal products. When the three craftsmen used old materials to create again, they were more dedicated to craftsmanship, rather than using it more functionally in the lives of users, I think this is an issue to consider. In addition to its aesthetic value, an original material will have more possibilities in our uses of life and this will give it multiple values. I think that when it really enters human life, when it has an impact on the human social environment and on environmental values, it is the way to truly recharge one's batteries.

3. Wool felt craftsmen shall obtain better conditions

A material alone cannot be transformed into a product, craftsmen or designers are the main means of its development. There are many processes for processing wool felts from wool to products that can be sold. During investigation, I understood that it was necessary to have certain provisions and conditions as a craftsman to carry out these processes.

First of all, perseverance, innovation and desire are essential. In our time, with the rapid development of science, technology and countless new materials, much perception and perseverance are needed to allow the return of an old material and know – how, long forgotten.

Secondly, craftsmen must have a cognitive ability to material properties. A complete understanding of the physical properties of material is a very important aspect of creation and design, both for craftsmen and designers who, if they do not know enough about the characteristics of the materials used, cannot get the most out of them. According to the craftsman **Li Qiang**, whose main customers are the elderly and patients from the nomadic region, his consumers, for geographical and cultural reasons, are very familiar with the characteristics of felts and can use this material to obtain the expected service or soothing, but if the designer does not understand the properties of the material, to some extent it will be limited.

Economic conditions are the basic conditions for the personal development or entrepreneurial process of the wool felt industry. Although there is a geographical advantage and the acquisition of raw materials allows cost savings, there are still many development risks and time will be needed to stabilize the situation. Economic fundamentals are also very important. Among the three craftsmen who participated in my research, **Ulantoyaa** was able to create a wool felt manufacturing company, not only because she worked as a photographer, but also because she opened a nomadic shop before building her company; these first experiences made it possible to obtain the basic financing to satisfy her entrepreneurial spirit. One of the main reasons for **Li Qiang's** limited development is his market and capital problems, which make him more inclined to private personalization in the manufacture of his products than to create and promote his own design products actively enough. Although Chinese government has put in place fi-

MARKET SURVEY

nancial support policies for artisanal heirs, the emergence of a large number of artisans does not allow all problems to be solved in time and these measures do not apply to all artisans. I think that, to a certain extent, most craftsmen are more attached to the demand for the honorary title of craftsman, and ignore the direction of the product itself. So I think that if craftsmen and designers want to develop the wool felt industry, they must be given the right economic conditions, which will not only reduce their manufacturing time, but also give them the confidence to focus on their products.

Impact of national policies. For the revival of arts and crafts, Chinese government has put in place supportive policies. For example, it has provided financial incentives for the qualification of craftsmen and supported a number of competitions in related industries, promoted crafts through the media, which has had a great impact and a large number of people have joined the industry. According to **Li Qiang**, most people involved in the manufacture of wool felt belong to the bottom of the society and do not have the necessary funds, the situation is difficult despite much motivation. By encouraging and assisting the qualification of craftspeople, government measures, in keeping with the spirit, do not have the same effect according to the social situations of each individual and are still too limited for some.

It is undeniable that national policy has a very important influence on the development of wool felt and that craftsmen will have more power in the industry, but it cannot be denied that there are still problems. A real progress can be observed in government policy to develop the product, but more investment should be made, technical and environmental qualities of this material should be promoted to obtain more convincing results and have a real impact on the wool felt industry.

National cultural and linguistic environment.

Wool felt, this ancient material, is produced mainly in nomadic areas. In China, for example, as with my research, most nomadic populations are mainly based in Mongolia and most of them are Mongolian. Mongols have their own language and culture. Many craftsmen in nomadic areas have a low level of Mandarin Chinese and many of them have their own method of making felt. If one of the parties does not speak Mongolian or Mandarin Chinese, communication is difficult. For example, artisan **Li Qiang** mentioned that his products are mainly from private marketing and that it is difficult for him to find a partner, especially because his region is an autonomous Mongolian region, but he does not speak Mongolian. While craftsman **Ulantoyaa** is Mongolian and masters Mandarin Chinese, the scope of cooperative products will therefore be broader. Students of the schools taught by the **Dalai Qiqige** craftsman are also Mongolian. In terms of language, **Ulantoyaa** and **Dalaiqeqeg**, who master both Mongolian and Mandarin Chinese, have a development advantage over **Li Qiang**. Craftsmen and designers are limited in the development of their products when they cannot integrate into the local cultural environment and language.

In addition, I think there are still other more important points. For example, in my communications with three craftspeople, I learned that they had not received any study or artistic training and that the history of wool was not known to them. They were unaware that Western countries had developed an art of wool felt and that the existence of design development was relatively unknown to them. On the one hand, from a historical point of view, industrial design is lagging far behind in modern Chinese history, on the other hand, most economically developed coastal cities have a population whose artistic thinking and awareness are relatively lagging behind, where product creation is mainly artisanal without taking into account the real values of the products.

In addition, attractive tourist areas have a wealthier population that is primarily aimed at the private sector, and there is no longer any reflection and investment in the functionality of the product or products with a modern meaning. I think that a professional craftsman or designer must do everything possible to understand the history of material, the source they use, its development, its performance and so on, because the clearer the use of their material, the clearer the learning and thinking and the open horizons to earn more and for more creativity.

4. Development wool felt craftmen from Inner Mongolia

L When he set up his workshop, all products of the craftsman **Li Qiang** were made by himself, sometimes with the help of his family members, but he had no other partners, his production model being relatively simple. Sales of its commodities are made either through the Internet or through other local people.

Ulantoyaa created the company at a time when she had a certain economic income. She owns the wool felt business and started the company as soon as she had sufficient economic income. Products are manufactured by the company's employees and they can sometimes take them home to process them. She appreciated characteristics of the inhabitants of the nomadic areas who could no longer find work after entering the city and recruited a large number of unemployed people, trained them in handicrafts and reached about 100 people. This is a good strategy to solve the employment problems of many nomads while developing the company. In the course of her work, she has actively conducted market research and sought opportunities for cooperation on many aspects. Currently, most of her partners are creators of interior handicrafts and sellers of tourist handicrafts. In addition to actively participating in local fairs and actively promoting her products, her sales methods are mainly based on cooperation with other companies and private customers.

Dalaiqeqeqeg even if she has not yet worked for long in her main company, has not had many economic and life difficulties during her career, she has tried to create a large number of different products and accumulated a lot of experience in creation. After wool felt became her main activity, she actively applied for a qualification as a craft heritage and participated in various wool – related competitions, while promoting her own products. She already has some influence in the wool felt industry.

We can see that **Ulantoyaa** has been in the industry for 8 years and she works in this field is the shortest among them, but her wool felt company is currently the most developed, her economic base, her rich experience, and her broad vision is important in her development.

5. The current state of the wool felt industry in Inner Mongolia

From a market perspective, finding a larger market is one of the most pressing needs of these trades. Many factors influence the market: for example, most current consumer groups do not use original materials. One of the conditions that influence the return of wool felt is how to ensure that the material is understood and accepted. If the material cannot really be recognized by people, it will only give them a certain "retro freshness" in the short term, then will remain at the level of craftsmanship whose demand will be especially high. Penny Spark, a British design historian, mentioned in the book "Introduction to Design and Culture" that, although technology has created materials, itself does not guarantee the use or acceptance of market materials. Designers play a key role in evaluating popular tastes and activities as well as in transforming materials required. Often, a designer plays a very important role: he must be able to use materials and ensure that the product affects the applicant, rather than being influenced by the applicant.

In a conversation with the craftsman **Li Qiang**, he mentioned an interesting phenomenon, saying that in his father's time, it was a time when craftsmen created what everyone used. In his private order, a Beijing customer had a pair of wool felt boots customized, but when he received it, he was not satisfied with the merchandise and thought that the thickness of the wool exceeded his thoughts, while the craftsman **Li Qiang** thought he had made the thickness of the wool felt most reasonable in terms of temperature protection. He then modified the thickness according to the customer's requirements. When the designer and the user have different types of thoughts, when the angle of attention is different, should the designer insist on his sovereignty? Who decides on the direction of the product? I think that the way to reduce the difference between the two parties is a favourable condition for products to be better marketed. So, whether it is a craftsman or a future designer, both must know more about the market and needs, and must also know each other well.

For the future of wool felt products, I believe that the reconstruction of the practical culture is very necessary. Maximizing the value of natural materials is an objective that deserves to be shared with all. To give the product a cultural history and content, which make it a creation with a history with a soul of warmth. The famous Japanese architect Tadao Ando once said that the ultimate goal of design should not be limited to "human use". He called on everyone to pay attention to the dialogue between industry and nature and to respect the original value of materials. For example, the yurts used by nomads need only a small amount of wood, wool felt and a few pieces of leather. Since ancient times, its invention and use have minimized the consumption of natural resources in the construction of this habitat. From an environmental point of view, it should now be recognized as the most environmentally friendly home in the world. Some developed countries have started ordering yurts in Inner Mongolia, which shows recognition of the product's environmental culture. So how do we deal with the relationship between people and the environment, materials and design? This is a problem that craftsmen and designers will have to face. We cannot say that we want to be environmentally friendly and on the other hand create pollution. In the process of returning to old materials, avoiding a large number of storytelling products, we should use relevant laws to standardize the wool felt industry.

For artisans in the Inner Mongolia region, not only do they want more people to engage in wool felt crafting, but they hope that more people will be able to track and use wool felt products. They expect that on the basis of the preservation of the craft heritage, the product will be pushed towards a larger market. How to transform cheap and ecological materials, give them more aesthetics, how to make materials that reflect the culture and values of another era, so that they have more possibilities today? This is what craftsman and designer must be thinking about.

The way artisans can transform themselves into modern designers, and if more designers can participate and pay attention to this industry, becomes the issues of today and the directions of future development. Whether it is a traditional craftsman or a modern designer, they must not let themselves go into a passive state. Before creating, you must have your own criteria and options for the natural environment, economic interests, social development and establish the most necessary environmental responsibility at that time.

If skilled craftsmen and art universities work together, more young people can be involved, making the material more aware and creating more modernisation from multiple perspectives. French designer Agnès Zamboni said: "Many design students now believe that they are not responsible for the choice of materials and manufacturing processes in product design. But in fact, they understand that they need to know this field, as well as the materials and processing methods, which have a decisive role in design. If designers do not know some of the existing professions, their creative potential will be limited, such as a composer who wants to write a complete symphony but cannot grasp the sound of the instrument." The future designer can be a person who combines many qualifications, including artists, engineers, materials experts, psychologists, human and environmental sociologists, philosophers, etc.

CONCLUSION

By investigating three wool felt craftsmen, we can understand some of the problems encountered by this activity in Inner Mongolia as well as the current situation of craftsmen involved in this industry. First of all, they are impatient and hope that the craft of wool felt can be passed on and accepted by a greater number of people. Secondly, they hope to find a larger market and that wool felt objects will be better used in everyday life. I believe that the concept of "heritage", on which the craftsmen insist, must not be confined to an act of tradition and copying. If traditional craftsmanship is to enter the market, it must take the initiative to adapt to the changes and needs of the modern market, rather than just sticking to tradition. We should achieve a more modern design based on the preservation of traditional cultural and design concepts. To give old wool felt materials a more modern design, modern craftsmen and designers must work together. By reintegrating wool felt materials into people's lives, more work will be needed to improve the industry's system, strengthen laws and regulations to protect a brand image and avoid the phenomenon of counterfeiting.

With social changes, reduction of nomadic populations makes it impossible to use wool felt materials as often as in the past. Faced with the long – term stagnation of old materials, we must now solve the problem of adapting to social change and the market. I think that in the design of wool felt products, we can materially use the explicit value of wool felt and fully reflect its hidden value. Craftsmen and designers need to know more and improve by having a complete understanding of the material. It is not possible to limit oneself to a single field: it is necessary to study several fields and learn to manage the relationship between materials and design.

The European Arts & Crafts movement shows that there is a successful heritage relationship between crafts and modern design, but the Arts & Crafts movement was mainly initiated by artists and designers, thinkers, etc., while the Chinese crafts revival movement is more government – led. However, after the beginning of China's reform and opening, the diverse artistic atmosphere in the West gradually spread and began to have an impact on it. Now is the time for a craftsman to transform himself into a modern designer and it is also the time for modern designers to participate in the joint creation with craftsmen. There should be more interaction between craftsmen and designers and the process of learning and interaction between them will also be the process of mutual transformation between the two. In the revival movement of Chinese craftsmanship, more young people will participate in the future and the future development of Chinese craftsmanship will be a reminder and a complement to the future history of global design.

Sustainability is extremely important for today's society: as a designer, we

CONCLUSION

must have a fundamental understanding and judgment about the choice of materials. Designers need more choices to determine the direction of design, they should design beyond a very commercial perspective, to put environmental protection as a prerequisite for designing more environmentally friendly choices with natural materials, to make the greatest efforts to achieve so that the environment creates minimal stress. As the Finnish designer Johannes Ekholm said, design must break with the concept of purely formal development: design must be a process of knowledge and recognition of things. Design must not remain on the surface, it must be integrated into life to make life easier. If the cultural concept of environmental protection can be conveyed through design, it will make the product more moving and make the design more meaningful. We should also let more people know the value of wool felt materials. How to maximize the value of this natural material and redefine its role is a goal worthy of our joint efforts. How can we get him back faster? A market that rebuilds user confidence is the next direction to consider. The struggle for the rehabilitation of old materials such as wool felt is also that of other materials that have disappeared or are still asleep; they are waiting for the designer to give them a new role.

APPENDIX

Li Qiang Interview

Date: 21 February 2018

Location: Sunite Right Banner, Inner Mongolia, China

Bao Aruna: *Hi, we are now starting an interview on your experience of creating in wool felt. Can you tell me about your basic situation, such as your study experience, your current professional situation, etc.? Thank you.*

Li Qiang: I'm a wool felt craftsman of Han nationality. After graduating from the technical school in Saihan Tara, Inner Mongolia, I have been working on this subject for about 21 years and have set up my own studio called Shunet

Felt. My house is next to my studio.

Bao Aruna: *Have you ever had a design learning experience or a secondary activity?*

Li Qiang: I have no secondary activity, I have never studied design and I started making wool after graduation.

Bao Aruna: *For which reason did you chose to do so after graduation?*

Li Qiang: The main reason is that it is difficult to find another job just after graduation. After graduating from the school of technical engineers, I only had a few mechanical assembly tools. I started making wool in 1995, it was taught to me by my father, because he was employed in a wool factory. I did it originally to live, then it gradually evolved into a pleasure. Now, after the promotion of public policy, I have more responsibility and motivation after I applied for the wool know – how. My products are handmade, which are characterized by a specific training of wool felt, felt and pattern can be made at the same time.

Bao Aruna: *What do you think about this industry now?*

Li Qiang: *I hope that someone can continue to do so in the future, that this good material will not be abandoned. I also hope that more people will need it and really understand its benefits.*

Bao Aruna: *You choose to do it, do you have the expected result?*

Li Qiang: I didn't think much about it at first. Mainly because the status of artisans in society is currently not high, most are poor. People in this occupation must initially face a situation in which life cannot be improved. If you can't make money, you won't have the motivation. But now, the government has given us help and honour through its policy. This emotionally creates real encouragement. Later, there were also art majors who came to learn from me. One, then created her own studio and became a designer. But I think that at the base are the conditions of the student's family, if they have entrepreneurial funds, it is a prerequisite for a bold project.

Bao Aruna: *What is the entrepreneurial effect for this student?*

Li Qiang: I think she is doing very well, but her studio is not only about wool felt, there are many other things.

Bao Aruna: *Is the current working condition a team, a person or a company?*

Li Qiang: *Mainly the production is independent and personal, sometimes family members come to help me.*

APPENDIX

Bao Aruna: *Why have you chosen to create your own studio? Is it a personal hobby?*

Li Qiang: I think that I am one of the few craftsmen to be in this field, other people do not know much about it. After the reform of the national system, most felt factories closed their doors. I am the only one in the region who has persisted. I thought about finding a partner, but I couldn't find the right one. Many young people do not want to work in this industry, most of those who know it are people over 70 years of age. Due to age, physical forces decrease and they can no longer do heavy work.

Bao Aruna: *Do you think there is a difference between you and other wool felt craftsmen?*

Li Qiang: For the few other craftsmen in this field, materials are mainly semi – finished products, they buy wool made by machine in the southern factory and are confined to processing. However, my wool felt is my own raw material and I have processed it by hand, in addition to a small number of private customers, who themselves supply the raw material in wool.

Bao Aruna: *Do you have any ideas about your work, your satisfaction, etc.*

Li Qiang: I think that a big problem is that nowadays, people don't like their own production, including some nomadic people, for example: this material, if the manufacturers themselves no longer use it, if it no longer attracts you, how can you develop it? It is only when designer or craftsman loves his own profession that he can be responsible for others. I think both people and government have problems, people do not cherish this material and the government pays little attention to it. For example, sheepskin, local elders cherish it because it was precious in the past, but one day I saw people throw it out on the street and some people burned it. Now, its value is one yuan, a piece of leather (about 0.2 euros). This problem would not occur if the previous fur factory still existed. So let others accept the premise of this material, you must first love it yourself.

Bao Aruna: *Do you currently have fixed customers or have you worked with other companies?*

Li Qiang: I started as a nomad in the local pastoral area. Later, after the promotion of the product, other customers also asked me to customize the finished product. But for pastoral areas, I mainly have some elderly people with leg problems, because they know the advantages of this material. At the same time, I also advertised on Internet. Now, the pasture is degraded and many shepherds have entered the city. People's lifestyles have changed, fewer people are using this material and the market is small. I think that if I only use it in pastoral areas, my sales will be very limited, but I don't know how to change.

Bao Aruna: *Do you know the historical source of wool felt? Do you have information about wool felt products in some developed Western countries?*

Li Qiang: I do not know much about the history of wool felt. I don't know much about Western countries, but I have participated in local exhibitions. The government has also sent people to investigate here, but few. I have tried to find relevant information on the Internet and asked others, but my own level of computer skills is not high. I don't have the best ways to understand it; the main source of my knowledge is that of my parents. Apart from the daily understanding of the material, wool felt has not changed for thousands of years and is stagnant.

Bao Aruna: *Can you give us an overview of some of the advantages and disadvantages of this material?*

Li Qiang: First of all, it can stay warm and breathable. The breathability of many pants is not very good, but this time it is different. It is hygroscopic and can be used as a mop, but its disadvantage is that it is easy to drop the wool. Wool has no harmful effects on the environment. It is soundproof and shock-resistant like some people who used to wear felt boots. Jumping a two-metre high obstacle is not dangerous, the felt has a good buffering effect. It is very flexible and can be molded at the same time. The material is as simple, price is low and the plasticity is practical. Its disadvantage is that it is easy not to take care and if it is not well preserved, it will be favourable to the development of insects related to humidity and temperature. Pure wool is a protein and like human hair, it must be protected. But we refuse to add chemicals such as pesticides.

Bao Aruna: *What are the specificities of wool felt compared to other fabrics?*

Li Qiang: First of all, I think the ancient material of sheep has inherited a spirit to some extent. There are many fabrics in the industry that bear the same name of felt, but most are not pure wool, or are of low purity by adding many chemical fibers. So if you use them to make standard items, such as yurts, chemical markers are neither practical, nor hot, nor very standard, but the price will be lower. Some people use it in the summer, for its low cost, but most tourist attractions look more like disposable items. Now, companies are more concerned about profitability and ephemeral users are not concerned about real or dummy wool felt. I think that handmade wool felt has an advantage in addition to protecting the environment, it is unique in its kind, it also has a certain collection value.

Bao Aruna: *Do you think you should have other conditions to do this job? For example, do you think the experience of arts education is important?*

Li Qiang: I believe that we must first work hard, which is inseparable from human personality, perseverance and physical strength. Secondly, I also know that knowledge of wool felt will establish the level of knowledge of materials, I think these skills are higher than those of art and that there is no experience in arts education. Talent is more important, you have to be innovative and try hard. I think it is useful to understand religion, for example, I have a customer who is a Miao in the south, a cow, a solid material and a more original sense, specialized in the customization of its objects. At present, national culture will play a leading role and the development of material has some advantages. I think that age will also have an impact, but more importantly, I love this material and I respect it. The basic economic conditions are also important, by shortening the manufacturing time, craftsmen will focus more on their own designs. However, I think it is even more important that the political support of the State should focus on increasing propaganda for crafts and strengthening the guarantee of venture capital, because design can currently only solve basic life problems such as food and simple clothing. The local environment is also important: although we have raw materials there, our city is small, far from developed coastal cities and the amount of information is limited. The advantage is that this will solve the problem of employment of some workers in nomadic areas. I often think that German industry is very famous, the quality of the products has remained unchanged for a hundred years and remains very powerful. I should be more focused and insist on doing this, just doing this.

Bao Aruna: *Can you share a work experience?*

Li Qiang: I have a deep impression: I have a customer who ordered wool felt for an ordinary temple, regardless of price: the only condition had to be pure white and the quality had to be good. Wool felt therefore also has a certain influence and status in religious belief. For example, when many people in the south saw the wool felt, especially the pattern above, they all knew it was a nomadic product, but the deeper nomadic cultural connotations were not known, so if a product loses its cultural taste, it changed the smell of the product and made it lighter.

Bao Aruna: *According to you, what is the greatest success of your product?*

Li Qiang: My first private customer was from another province, but he lived very young, in a nomadic region, he felt nostalgia when he found me and wanted to have a pair of nomadic felt boots made. After that, advertising was opened and some people began to want to learn this profession.

Bao Aruna: *What kind of wool felt do you and can you briefly present them?*

Li Qiang: I mainly use felt to make boots, hats, 12 zodiacal art felt paintings, cat cabins, sheep bags, computer bags, etc.

Bao Aruna: *Did you try another design work?*

Li Qiang: Fineness and thickness of the wool felt make the product different. I have also tried extensions of products for the home, such as the tablecloth, which is usually slippery, but if you put on a little felt, it will no longer slip. However, the main problem at the moment is that there is not enough time, there is not enough manpower and many materials on the market have progressed while the felts have stagnated, they have not changed much. I have a language problem mostly, I only speak Chinese and I don't understand Mongolian, some Mongolian students are curious and interested, but it is difficult to communicate a common language. I think it has a linguistic advantage for Mongolians. and people are easier to accept it.

Bao Aruna: *Do you have any ideas on the current development?*

Li Qiang: I think it's mainly a market problem, there's no big market for this material, so I essentially expect customers to come and order custom products from me. I do not lack raw materials, there is a lot of wool in the pastoral area, if it can be used to make more valuable products than wool waste, making a good quality will not be a problem for a hundred years.

Bao Aruna: *Have you ever thought about working with other companies or industries?*

Li Qiang: I have thought about it, but I don't have any concrete practice, the main reason being that most people do their own work, most of them are profit-oriented and have little interest in old materials. I didn't work with a designer, but a design student came to me to learn how to do it.

Bao Aruna: *Do you take information about people's needs before designing a product? Have you done any research on users, etc. ?*

Li Qiang: I have not studied it, I don't think about it much, I often do it before I think about it, I put it on the Internet and ask others to ask questions. Most of the time, it is private personalization, I prefer this formula, others provide ideas for me.

Bao Aruna: *What types of customers are looking for your customized products?*

Li Qiang: People who are in great need of this material, who generally need felt boots and felt for other uses. The winter insulation effect is good, the small batch, the cost of wool is low, but the manual cost will be higher. Another type of customer is rich, he orders products and demands high quality products.

Bao Aruna: *Do you think there is a limit to this industry now?*

Li Qiang: First of all, the staff is limited, some people from Xinjiang come to my house to make more than 100 products, but I am alone and my work is not enough, it is difficult to have more than creation. Many people now think that wool felt is too old, not fashionable and very limited. People who need this material are for some, because of their attachments, others are special customers, such as those who suffer from leg pain, back pain, diabetics also need it. In addition to insufficient publicity, with the exception of low demand in other cold regions, people, especially young people, do not understand woolen felts well, they do not know what they can do with them.

Bao Aruna: *Why do you think there are not many people who choose to make felt from whool?*

Li Qiang: Firstly because of people's thoughts about this material, I don't know, there are also many who reject old things and feel that they don't get any economic benefits. For example, in summer, the quality of the wool is good, but the summer is very hot, few are willing to work it and in winter, the production cost is relatively high.

Bao Aruna: *Have you ever thought about giving up this industry?*

Li Qiang: I thought about giving up before, but since that's what my father told me, I'm responsible for the inheritance. Later, after the government supports it, we will have honours and subsidies for ourselves. We had nothing before and now the situation is improving.

Bao Aruna: *Have you made any changes since the creation of the studio?*

Li Qiang: The quality of the product will be improved: once the quality has improved, the requirements will be higher, the price of the product will also increase and with the development of the company, the user will need beauty, comfort and pay the price of the product. For example, in the past, farmers focused only on warmth and robustness; with the improvement in living standards, there are more demands on aesthetics and comfort. The wool is very hard and not easily deformable, but now people prefer to use lambswool, which is more comfortable. When the user's request is improved, the product will be classified, more standardized and the finish will be more refined; all these elements will be linked and the cycle will be improved.

Bao Aruna: *You mentioned that some companies make wool felts made by machine, are they different?*

Li Qiang: According to my experience, handmade wool felt has a good density and high quality, but it takes a lot of time and low efficiency is not good for mass production. But the wool felt produced by the machine is not dense enough, the hardness is low, the wool is easy to break, the yield is high, mass production and marketing is easier. The wool is sent to the southern factory for processing, but the freight is also very high. Many factories use dry wool and wool waste that nomads do not want; these wools are mixed with other impurities, such as chemical fibres, so that the quality of the wool felt produced by the machine is not good and does not contribute to the protection of the environment.

Bao Aruna: *Are you in a relatively passive mood now?*

Li Qiang: Yes, now I'm really passive, because people have no idea of the right materials: for example, in the past, every household had boots or wool felt clothing, just like raincoats.

Bao Aruna: *More than what you mentioned, there are currently few people who know this material. Are there other differences that affect the development of design between craftsmen and consumers?*

Li Qiang: There are differences: for example, many people live in different geographical environments, the perception of cold and heat is different and everyone's perspective will be different. For example, there is a customer in Beijing who booked a pair of wool felt boots from a network photo but, after upon reception, he finds it too big and too thick: the craftsman has made his environment in the most suitable thickness.

Bao Aruna: *Have you ever thought about applying for a patent, marketing the product and protecting it?*

Li Qiang: The name of my studio is "Shunet", in reference to our nomadic region, but written differently. "Su" represents comfort and "ni" represents the finesse of the wool, which means quality. But I didn't think to apply for a patent and that applying for a patent would limit product development and creation, but I hope more people will be able to do so.

Bao Aruna: *Do you have more relevant industry guidelines?*

Li Qiang: There is no relevant standard, I mainly insist on handcrafting. The current design is customer - based, what the customer wants to do. First, meet the needs of the customers and then take time to do the thing. The colour will not only be pure colour, but also other dyes according to the customer's requirements.

Bao Aruna: *About product promotion, what are the ways?*

Li Qiang: Local media had interviewed the newspaper and promoted it to the Mongolian school to do activities, but my language presents some obstacles, I participated in some fairs and also sold on the Internet. But I think that the promotion of wool felt materials is more important than the promotion of products: for example, advantages of wool can only really develop if more people can understand its characteristics.

Bao Aruna: *Do you know that there is a difference between traditional craftsmen and designers?*

Li Qiang: I don't really know the difference between them, but from my point of view, although I haven't studied art, the role of wool is above all physical and industrial experience and my technology will be more solid. The design part of the wool felt can be done by people who need an artistic experience, for example designers. Times are different now: In the past, manufacturers influenced the applicants, but now applicant influences the manufacturer. What he needs is what we create.

Ulantoyaa Interview

Date: 23 February 2018

Location: Saihantal, Inner Mongolia, China

Bao Aruna: *Hello, we are now starting an interview on your experience of creating in wool felt. Can you tell me about your basic situation, such as your study experience, your current professional situation, etc. ? Thank you.*

Ulantoyaa: After my bachelor's degree in Mongolian studies, at the Normal University of Inner Mongolia, I worked as a photographer in Erlian City from 1992 to 1995. Later, from 1995 to 2010, I opened a Mongolian style craft shop. In my store, I exhibited wool felt crafts. Then, from 2010, I started a business with my husband to become a boss and I am now a wool felt craftsman.

Bao Aruna: *Why did you chose to make wool felt craftsmanship?*

Ulantoyaa: When I opened the ethnic products store, I was exposed to this profession. My father loved animals and when I was young, I was in contact with animals. My godmother was a famous folk craftsman, I was influenced by them.

Bao Aruna: *What do you think of this industry now?*

Ulantoyaa: My biggest concern now is how to communicate with people: how to make them understand that the wool felt industry is very important. Since wool felt has been abandoned for many years, many young people do not understand it when it reappears. Most people only know that wool felt is a craft and do not know its culture. For example, embroidery with camel hair on wool felt is a custom of this material. Many young people do not know the role of wool felt: it fulfils many functions: it can be widely used in life, as in the manufacture of a mattress. It can also be used for medical treatments, burning pieces of wool felt and pressing the wound to stop bleeding, in some places cotton is used to treat. In the ancient wars, it was also used as protective clothing. People use flexible characteristics of wool felt to put high – density wool in armor to protect the skin, help in earthquakes, counter moisture and mitigate the offensive weapons. Wool felt is breathable, non – polluting and warm, and clothing is manufactured according to these characteristics. It is also used in industry for sound insulation. In short, wool felt has many advantages.

Bao Aruna: *Can you tell us about your current type of work?*

APPENDIX

Ulantoyaa: My husband and I have created a company together, the company is based on the cultural industry of our canton. The company now has about 100 employees. At first, the company had only a few, but the number of products we could produce was too small to be commercialized. Later, I noticed that many nomads could not find a job after joining the city, there were many unemployed, especially those living in the city. So we started trying to hire these people and teach them handicrafts, because many of them came from nomadic areas and they knew wool felt well. Gradually, more and more people are becoming more and more qualified.

Bao Aruna: *Do you have any other ideas about your professional status?*

Ulantoyaa: I believe that market development is currently a problem.

Then, make a big product or a small product, what is better? What aspects of life can be applied to products, these are some of the problems I am currently facing.

Bao Aruna: *Have you already carried out a market study on the needs of product users, for example what kind of people may need this material?*

Ulantoyaa: Yes, I have been to many places, I bring my own work to present it to others, so that everyone knows that the product is ecological and I think that in the end there will be more market. Many people ask me about the benefits of this wool felt for the body and what it does, but it is mostly about the use of design. Most people don't usually pay attention to drawings. I present more explanations on the motifs and cultural characteristics of felt, because I have knowledge of crafts. I have been to some of China's main provinces, Hebei, Henan, Jiangsu, Guangdong, etc., most of them are located in tourist sites where there are craft collections. I find people interested in what nomads do themselves, because they think the material is real, unlike machine made products. Just like mentioning that a certain dairy product comes from Inner Mongolia, people think it's good because there are grasslands there. For example, in nomadic areas, meat is eaten all year round, the habit of cooking meat is different from that of the south, the method of making wool is different and the effect is also different. I also went to Hohhot for research and made wool felt for the interior decoration of many Mongolian restaurants, which require more typical cultural products.

Bao Aruna: *Have you seen wool felt products in other market studies?*

Ulantoyaa: When I was doing research, I saw Russian and Mongolian products on the wool felt market and I discovered that the products manufactured contained different types of thin and thick fibres, for bags, clothes, shoes and hats. As our traditional wool felt is generally quite thick, it has been found that the same manufacturing process can be used to adjust hardness and density and that many types of designs can be made. I also noticed that they also had a certain design for the dressing of certain household objects, such as sofas, electrical protective covers, but most of them are not made of pure wool, they are mixed with other chemical fibres. I also noticed that some areas of fashion are now popular thanks to wool felt clothing and that the price is very high.

Does Bao Aruna: *Does you company currently have fixed customers or have you worked with other companies?*

Ulantoyaa: My company is called Tengli National Culture Communication Co., Ltd. I have permanent customers. For example, there is a yurt production company near our city, which is a very well – known production company. It takes a lot of wool to make the yurt and many are from me. We have a fixed business relationship. He and I grew up together, he only made yurts, we can just cooperate. Wool, felt making is a tradition of our nomads, so I am more skilled. Sometimes I also work with other companies, such as interior design companies.

Bao Aruna: *Do you have historical sources of wool felt and information on wool felt products in Western developed countries?*

Ulantoyaa: I do not know the history and culture of wool felt, but there is a 70 – years – old wool craftsman from Mongolia who said: An ancient man mounted on a horse was discovered in a tomb 3000 years ago. The clothes he was wearing were made of wool felt. We can therefore judge that there are 3000 years of history of wool felt.

Bao Aruna: *Do you know its production process?*

Ulantoyaa: The traditional procedure for Mongolian wool felts is probably the same: they cut the wool in May and June and start making wool in August and September. When they made felted wool, they chose the right day, usually after drinking tea with milk, could not let the children cry on that day and could not make an old man angry. They spread the wool on the table, sprinkled milk and water on it, rolled it and had it trodden by horses. I found a video previously made on the web. They would leave some wool and save it to make something the following year. They thought it was like winemaking and that it took time to improve quality. Wool felt should not be placed randomly and should be carefully stored. An important characteristic is that the wool felt used in the yurt skylight must use lamb's wool, the most beloved, or the young sheep is considered the most spiritual. There were always special thoughts for all the wildlife of nature: they took good care of their children and were particularly good for their parents. The use of their wool in traditional culture, mainly expresses the love for animals and respect for nature.

Bao Aruna: *Do you encounter any difficulties in making felt?*

Ulantoyaa: There are no particular difficulties, everyone helps each other, some people have never received cultural education in the nomadic region which sometimes leads to communication difficulties.

APPENDIX

Bao Aruna: *What are the disadvantages of wool felt?*

Ulantoyaa: The main problem with wool felts is that it is easy to neglect wool, but we refuse to use chemicals, because we believe that it is not natural and would affect the most natural environmental conditions.

Bao Aruna: *Have you compared with other fabrics?*

Ulantoyaa: I did not compare it because I never thought of replacing it with other materials.

Bao Aruna: *Do you believe there is a need of development conditions to continue this business.*

Ulantoyaa: In the past, some people came from Russia to import products, but the cost is much higher. I think we have a lot of advantages, why not do it ourselves. There are no longer many factories in the nomadic area: to protect the water source and the natural environment, most of our wool felts are semi-finished wool felts from industrially developed urban factories. We then process it manually. Wool cleaning does not affect the environment: most nomadic areas use trona (natural hydrated sodium carbonate, monoclinic) to wash wool, but our region belongs to the semi-arid prairie area and water sources are scarce. Therefore, if the demand is high, only wool felt produced in other regions can be used.

Bao Aruna: *Do you have any special considerations when choosing an employee?*

Ulantoyaa: When our company recruited employees, I wanted to know if they had already experienced the manufacture of wool felt or if their parents had made it. Some of my employees have worked in the felted wool sector, others have practiced a little through their leisure time, others want to do it for themselves but cannot find the market, there are also a large number of unemployed people from nomadic areas. Most people are interested in it and I think it is very important, so that it goes well in this work. There are also others who wish to do so, but they do not have economic conditions and they can only come to me to be employees. Most of the former nomads made wool felt to meet their own needs, they rarely get sales on the market because they have no marketing skills. I think that's one of the reasons why this material is decreasing rapidly. We are now trying to produce a brand effect, so that the product, once its reputation has been established, can be brought to market more quickly.

Bao Aruna: *Can you share a work experience?*

Ulantoyaa: At the beginning, we presented our wool felt products in a pearl festival organized in our region. The pearl festival takes place every year on May 12 and during this exhibition, I met many friends and exchanged ideas for the joint production of wool felt. During this exhibition, I met my first customer and obtained the first order. Later, our work also caught everyone's attention; many people came to interview us and asked questions about how to make wool felt, including many university professors. From that, I discovered that there are still people who are very interested in wool felt crafts and that many people also like the crafts we make. Later, a Russian invited us to Russia to present the work and teach the technique, but I regret that he missed the opportunity because of other problems.

Bao Aruna: *Have you ever tried new innovations in your creation?*

Ulantoyaa: The fact that was confusing at first was to know which material to use to process wool felt. Wool felt is natural and ecological and if I use treated wool, it is chemically coloured, it does not correspond to my manufacturing idea. When I was a child, my father raised many camels. The camel neck hairs are different depending on the depth, I have divided them into more than 20 colours, which is ideal for the use of camel hair, keeping the originality and environmental protection of the product.

Bao Aruna: *Which kind of wool felt do you have, can you briefly introduce them?*

Ulantoyaa: Most of the time, I mainly make carpets and handmade felts, but wool felts are very useful for people. I used it to make wine bags, kitchen insulation cushions, costume hats, socks, bags, gloves and bow and arrow bags. In order to get more ideas, I also organized a competition with 80 participants, allowing them to propose a creative production in 6 days; at the end there were 38 types of objects. I will try to put them on the market in the future. During the production process, waste is produced and small waste will be used to produce small products, such as ornaments. We were only doing soles and now I have no ideas. However, at the moment, painting on handmade felt is the most widespread: many people think that wool felt is a craft and the tourist attraction is the main market.

Bao Aruna: *Do you have any ideas about current development, what are the limits, such as age, etc. ?*

Ulantoyaa: We all know that the price of craftsmanship is generally higher than that of machine manufacturing. Many of my employees came from pastoral areas and their horizons are not so open. Although I often encourage them to participate in the contest to broaden their horizons, many people are more limited in their expression. There are schools that invite me to class, but I mainly develop my activity around my company. I think that as long as there is interest, people of all ages can do it and there is no geographical restriction: when our employees are qualified, they sometimes go home to work.

Bao Aruna: *Have you made any changes since the creation of your company?*

Ulantoyaa: Even though the finition of my products and their technical maturity have improved considerably, the price has not changed. We now buy wool felt and then process the product by hand. I don't want the machine to replace the manual. One of the reasons I chose to buy wool felt made by the machine is that the felt washed by the big machine is cleaner, but the cost of the machine and freight are not cheap. If you consider touching it by hand, it is better to make the felt by hand, but the machine saves time and is beneficial for a large number of products. Therefore, I am mainly responsible for processing wool felt only by hand, in order to obtain the best quality of our products. There are many people who help me with advertising and I think it will get better and better.

Bao Aruna: *Do you think that national policy will affect your development?*

Ulantoyaa: First of all, the impact for me is there, but the national policy is limited in time: for example, the development time of our company's production has certain requirements. Even without political support, I believe that this industry can also develop: our raw materials all come from nomadic areas and we will not be limited by raw materials, we can do it as long as we want.

Bao Aruna: *How do you resolve different opinions about the product between you and the customer?*

Ulantoyaa: It depends on the situation and varies from one person to another. For example, when designing a hotel, it is necessary to take into account conditions such as aesthetics and ease of dismantling, which is different from home design. We are generally organized to meet their needs. Of course, we know more about materials and we will also be in conflict with design, which requires further exchanges and negotiations.

Bao Aruna: *How did you bring the product to the market?*

Ulantoyaa: We made design samples, then attracted customers with these samples and most of them placed orders after receipt.

Bao Aruna: *How can you promote the product?*

Ulantoyaa: Product promotion is done through various newspapers, interviews, television, internet, etc. I have also been to many exhibitions of traditional festivals.

Bao Aruna: *Do you think that there is a difference between traditional craftsmen and designers?*

Ulantoyaa: There must be a difference, I think craftsmen will know more about the culture, will understand the characteristics of the materials and their uses, but designers can have more ideas, I have never met them.

Bao Aruna: *Do you want to work with a designer?*

Ulantoyaa: I have worked with interior design companies and made small items for them, but I haven't approached other designers. If there are opportunities, I hope more people will try to do different things: since felt has many characteristics, it can soundproof, etc. I am currently in this industry not only to make money, but because I have a lot of empathy for it. Its nomadic culture was born from nature. Its role in protecting the environment between man and nature is the hope that this can continue.

Dalaiqeqeg Interview

Date: 25 February 2018

Location: Хянган аймаг, Inner Mongolia, China

Bao Aruna: *Hello, we are now starting an interview on your experience of creating in wool felt. Can you tell me about your basic situation, such as your educational experience, your current professional situation, etc.? Thank you.*

Dalaiqeqeg: After my Mongolian language degree from the University of Inner Mongolia, I worked as a teacher in a local college from 1983 to 2017. I worked felt wool as a secondary activity. After leaving school in 2017, wool felt craftsmanship became my main activity.

Bao Aruna: *Did you have you an artistic experience before?*

Dalaiqeqeg: The Know – how of the patterns on wool felt that I now use was taught to me by my mother when I was a child and I did not study art.

Bao Aruna: *For which reason did you choose to do this work?*

Dalaiqeqeg: I was mainly influenced by my mother, who was a wool felt craftsman. In the previous era, she raised several children by dressing them in wool felt to go to school until she continued to sew on the wool before her death. Because of my feelings for my mother, I hope to be able to continue with her expertise and I am personally very interested too.

Bao Aruna: *What do you expect from your commitment?*

Dalaiqeqeg: I didn't think much about it, but after making wool felt objects, many people were touched by my work. They learned that this trade existed. I am also very happy myself, because it is also the traditional culture of our nomadic people. It was about to disappear. I think it is a mission to continue this profession.

Bao Aruna: *Are you currently working as a team, a person or a company?*

Dalaiqeqeg: Mainly, I do it by hand and I also have my own studio. Sometimes my family comes to help me make wool felt. I also go to the surrounding colleges to promote the culture of wool felt crafts and teach them some craft techniques. I have more than 300 students to date and there are ten apprentices outside the school.

Does Bao Aruna: *Do you have studios or entrepreneurs among your apprentices?*

Dalaiqeqeg: Oh no! They are still studying at school, but when they have time, they will come here to study.

Bao Aruna: Do you currently have any ideas about the future of your work?

Dalaiqeqeg: I hope that I will be able to do more work in the future, that my products can have a collector's value, can be used as souvenirs, or for their use value etc. Therefore, I am currently trying to open up the market for wool felt crafts by publishing, participating in exhibitions and other activities to allow more people to understand wool culture and crafts. I also want to create a company in the future, the name has already been thought of, it will be called "Chagandalai", it also means the purity of wool felt.

Bao Aruna: *Do you currently have clear design objects?*

Dalaiqeqeg: There are also some, like the little felt boots with rabbit, many products are designed for children, it's more attractive, but now my studio is relatively small, the know – how in wool felt is relatively long to acquire. I still lack human resources.

Bao Aruna: *Do you know the historical source of wool felt and do you have information on the design of wool felt products in Western developed countries?*

Dalaiqeqeg: I don't know much about this story and I don't know much about Western countries.

Bao Aruna: *Can you give us an overview of some of the characteristics of this material?*

Dalaiqeqeg: The wool felt can be used in medical applications, it is shock, radiation and moisture resistant. Burning wool in the sheepfold in the pastoral area can dissipate unpleasant odours and purify the air, which is a bactericidal effect. High – density wool felt can be waterproof and when the waterproof is used, it is not easy to damage it, the felt from two or three thousand years ago has not rotted and has not deteriorated. It can be used for a long time and has the advantages of being warm in winter and cool in summer. However, if it is improperly stored in a damp place, the wool felt will be invaded by worms, so it should be discarded.

Bao Aruna: *Why don't you choose a machine to make wool felt?*

Dalaiqeqeg: The main point of my insistence on working with wool felt is the "handmade" which can protect this traditional craft. Our nomads began to make felt by hand as early as the 13th century and have used them in everyday life up to now. But now, the social environment has undergone enormous changes and this craft is about to disappear. On the one hand, it is the legacy of technology and it means that more people understand it, on the other hand, it is better in terms of product quality and the characteristics of wool felt.

Bao Aruna: *Do you find it difficult to make wool felt?*

Dalaiqeqeg: Wool felt manufacturing process is difficult, it requires not only qualified practice, but also consumes a lot of time and effort.

Bao Aruna: *What is the condition for doing this job?*

Dalaiqeqeg: I do not believe that there is any particular condition, regardless of age, gender or ethnicity; as long as people love it, they can persist. Economic conditions sometimes have an impact on its development and national policies can also have catalytic effects.

Bao Aruna: *Can you share experiences and ideas for work?*

Dalaiqeqeg: Before the death of my mother, although she did not go to school, she recorded in detail her knowledge of wool. I hope that when I make the wool felt, I can slowly accumulate my know – how and make a book out of it. After starting to make wool felt, I won second prize at the Xing'an League's second intangible cultural exhibition in 2013. Later, I won many awards in 2016 and 2017, with some popularity.

Bao Aruna: *What do you think about the current development of the product?*

Dalaiqeqeg: I hope that there will be a certain combination between product and culture, as for many crafts now, which was not the case before. For example, the bag of wool felt scissors I made corresponds to a taboo in Mongolian culture. You can't point the pointed end of the scissors at each other. I often have to use scissors, so I created a scissor bag and I hope that the cultural value of the product is greater than its appearance.

Bao Aruna: *Have you ever thought of working with other companies or industries?*

Dalaiqeqeg: I have thought about it: for example, some local car and furniture factories cooperate in the use of certain products, but I have just retired a year ago and plan to do so later. And I think my product has not yet reached the highest quality.

Bao Aruna: *Have you ever thought about people's needs, users' expectations, etc., before designing the product?*

Dalaiqeqeg: Not yet.

Bao Aruna: *How do you estimate the cost of the product?*

Dalaiqeqeg: The Product cost is not a problem, my brother lives in the pastoral area, the family raises sheep and wool is everywhere. But the quality of our wool is a little thicker and now the fine wool fabrics are imported from Australia, New Zealand or Mongolia. Their wool varieties are very fine and soft. The quality of the wool is related to the breed and region of the sheep.

Bao Aruna: *Do you know that there are many craftsmen who work with wool? Is there an age limit?*

Dalaiqeqeg: There are not many people who make wool in this area, but there are many grasslands nearby. There are relatively few nomads in our region and as the social environment changes, few people are familiar with this profession. There are also many young people who think that wool handicrafts are dated and do not like them. For example, there is a woman who wanted to bring together other women in her town to make wool felt crafts. She had studied with me, but she gave up after a while. I think first of all that many people have less cultural recognition of this traditional craft and secondly, they think that this work has not brought them sufficient economic benefits. Today, the company is growing rapidly and there are many types of products: many materials are replacing wool felt and consumers do not understand the benefits of this material. More importantly, when people choose products, they generally give priority to cheap products. This corresponds to a lack of social and environmental responsibility because there are many counterfeit materials in society. Although some people have already become aware of this problem, they are more willing to spend more money to buy products from abroad and feel more secure. China has started to take measures such as the revitalization of handicrafts, etc. So I think now is the right time, the key is how to promote the rebirth of the traditional culture of wool felt. I don't think there are many relationships between production and the age of the artisan, but the elderly are more experienced and do better. There are still regional cultural restrictions, as nomads have relied on cattle and sheep since ancient times to better understand them.

Bao Aruna: *Do you think there is a limit to this industry now?*

Dalaiqeqeg: I believe that craftsmen should be more open – minded and not allow wool felt to be limited to nomadic areas. We should consider other possibilities in other regions. While promoting and protecting the cultivation of wool felt, we can open up a wider market.

APPENDIX

Bao Aruna: *Have you made any changes to your initial production for the current product development?*

Dalaiqeqeg: First of all, the quality of my products has improved and it has become better than before. I'm just retiring, I'm still in the product development stage, there are no large-scale sales, but the demand is stronger than before.

Bao Aruna: *Do you have any ideas about industrial felt?*

Dalaiqeqeg: From what I understand, the industrial felt material is not good enough and will look like other fibres. The effect of pure, handmade wool is better and this material is ecological and durable.

Bao Aruna: *Are you saying it's stagnating now?*

Dalaiqeqeg: Even though the government has put in place policies that are conducive to development, personally, I do not rely much on policies. I pay more attention to the quality of my products and I persist in doing my best. This job, which is now part of my life, is part of my hobby.

Bao Aruna: *Is there a consensus about the product between you and the customer?*

Dalaiqeqeg: Usually, I introduce them to the characteristics of my products and their material advantages, but if they are not satisfied, I have to change my products and do what they want.

Bao Aruna: *Have you ever thought about applying for a patent, marketing and protecting the product?*

Dalaiqeqeg: I wish to register the name of the company "Chagandalai" in the future, which means a natural protection of the environment and I know that my products are pure, natural and ecological. After that, I will apply for a patent, register a trademark when the product is more mature and establish cooperation with others, if the conditions allow. I hope that my products are mainly woolen felts that give people a natural and environmentally friendly impression.

Bao Aruna: *Concerning promotion, what are the means?*

Dalaiqeqeg: Local media advertised and many friends helped me promote it, I taught this profession in the local Mongolian school, I participated in some exhibitions and promoted it on the Internet. But I think that the promotion of wool felt materials is more important than the promotion of products: for example, the advantages of wool can only really develop if more people can understand its characteristics.

Bao Aruna: *Do you think that there is a difference between traditional craftsmen and designers?*

Dalaiqeqeg: I'm not quite sure. I think that many young designers do not know much about this old material and that it does not attract them, their culture on this subject is even less clear. Many people think it is too backward to bring economic benefits. This requires a stronger sense of social responsibility, for example if a particular wool waste is not used, it will actually be a waste of resources. We just want to use it to realize its greatest value.

BIBLIOGRAPHY

Sources:

Interviews:

(1) An independent craftsman: Li Qiang (Chinese).

(2) An owner: Wulantuya (Mongolia).

(3) A teacher: Dalai qiqige (Mongolia).

Books:

(1) Buhechaolu, 《La culture de la maison Mongolie》, 01 janvier 2014.

(2) Penny Sparke, 《An Introduction To Design and Culture: 1900 to the Present》(Anglais), 06 décembre 2012.

(3) Agnes Zamboni, 《Matiéres et Design》(Français), Mars 2016.

(4) Shou Zhi Wang, 《A History Of Modern Design》, janvier 2001.

(5) Chris Lefteri, 《Materials for Design》(Anglais), Avril 2017.

(6) AMur Batu, 《Mengguzu Tuan》, Juin 2016.

(7) Jacques Morizot, et Roger Pouive, 《Dictionnaire d'esthétique et de philosophie de l'art》.

(8) Guo Yuqiao, 《Xi Shuo Meng Gu Bao》, 01 août 2010.

(9) Kenya Hara, 《Designing Design》(Japonais), September 2010.

(10) He Dong, 《Business & Economics》, 01 november 2009.

(11) Stéphanie CAUX 《La mongolie, une croissance, mais à quel prix?》 IRIS (Institut de Relations Internationales et Stratégiques) – Master 2 Geoéconomie et Intelligence Stratégique 2014.

WEBOGRAPHY

[1] http://www.sxnews.cn/p/64995.html.

[2] https://sns.91ddcc.com/t/50463.

[3] http://www.sxda.gov.cn/index.php?m = content&c = index&a = show&catid = 56&id = 12467.

[4] http://www.chnmuseum.cn/Default.aspx?TabId = 212&AntiqueLanguageID = 692&AspxAutoDetectCookieSupport = 1.

[5] http://www.bukhara-carpets.com/img/yurt/DSCN5853.jpg.

[6] https://www.flickr.com/photos/ninara/5868624139/in/album-72157624401124997/.

[7] https://www.pinterest.fr/pin/156851999492679973/.

[8] https://kids-magazine.com/woonya-petits-chats-a-adopter/.

[9] http://www.sohu.com/a/191283363_735370.

[10] https://www.pinterest.fr/pin/371476669233974681/?lp = true.

[11] https://www.centrepompidou.fr.

[12] http://www.romerovallejo.com/.

[13] http://www.metylos.com/.

[14] www.maxime-lapouille.com/.

[15] http://www.huntingandnarud.com/.

[16] https://www.youtube.com/watch?v=Hz5TAhZqqsY.

[17] http://www.journees-du-patrimoine.com/SITE/musee-feutre-mouzon-128656.htm.

[18] https://www.maudvernay-creations.com/laine.

[19] https://www.tendance-feutre.com/pages/Pourquoi-le-Feutre-de-Laine.

[20] https://www.airbnb.fr/rooms/489668.

[21] https://www.dailymotion.com/video/x75zra.

[22] http://www.lefeutre08.com/felt/feutre-laine.html.

[23] https://www.tendance-feutre.com/post/2017/07/23/l-origine-de-la-feutrine-%3A-une-histoire-qui-nous-vient-d-italie.

[24] http://lesonduchablon.blogspot.com/p/histoire-du-feutre.html.

[25] https://bellesrobesfeutrees.jimdo.com/la-histoire-de-feutre/.

[26] http://finance.ifeng.com/a/20140710/12688263_0.shtml.

[27] https://news.cfw.cn/v234930-1.htm.

[28] http://www.yltex.com/thread-25712-1-1.html.

[29] https://www.flickr.com/photos/ninara/5868624139/in/album-72157624401124997/.